Praise for
Defending the Constitution behind Enemy Lines

"Navy Commander Green expertly weaves historical analysis together with this incredible true story of a courageous fight for medical freedom."
—Robert F. Kennedy, Jr.

"As the son of a Navy officer and a convert to Catholicism through the AMS (Archdiocese of Military Services), I applaud Commander Robert Green Jr. for his courage and fortitude in taking a stand against the tyrannical policies and fear-based coercion of the United States government regarding the COVID-19 vaccines. A military is only as strong as its leaders and this book exposes the many injustices that were done by government officials to those who serve our country. Commander Green and all those in our military who resisted and were punished for not taking the experimental COVID vaccines are American heroes; it remains to be seen what the effects of the vaccines will have on our military personnel. This book is a testament to all true patriots who love God, country, and freedom!"
—Fr. Donald Calloway, MIC, STL; Vicar Provincial – Marian Fathers; author of *Consecration to St. Joseph*

"With the fight for traditional values and basic constitutional rights now at our doorsteps, *Defending the Constitution behind Enemy Lines* boldly steps into the culture war raging for the very heart of our nation. Not since before the founding of our nation have the American people endured such a long train of abuses and usurpations. Commander Robert A. Green Jr. represents a new generation of courageous military leaders who are willing to risk everything to lean into the fight against a runaway federal government and a politicized military leadership. Every freedom loving, God-fearing American must join in the effort to push back the tide of tyranny. We must pray, we must vote, and we must stand for truth!"
—Tony Perkins, president of Family Research Council and author of *No Fear: Real Stories of a Courageous New Generation Standing for Truth*

"This story is not only one of a critical shortage of moral courage at the highest levels in the United States military, but also one highlighting what everyday Americans who have dedicated their very lives to defending liberty are capable of doing in its defense. Commander Green is taking great risk by publishing this book, and future Americans will see him as a hero, sacrificing everything he holds dear for us, even those who seek to punish him. This book should be taught in every public school, every military basic training course, every military officer accession course, and have a place in every American family's home library. It is truly a 21st-century handbook for the American patriot."

—Rob Maness, Colonel USAF, (Ret.)

"Getting off the bus in the pouring rain at 0300 in late winter 1983, I was greeted by the echoing shouts of my name emanating from the battle-hardened Vietnam veteran who became my basic training drill instructor. After revealing to me that his best friend, who shared my name, was killed in Vietnam, Sergeant First Class Jessie Canales, gave my 18-year-old self the first military leadership position I would hold; leading our basic training infantry platoon. Since leading that platoon during the military rebuild after the decimation that followed the Vietnam War, through my time as a Green Beret, my service as a Special Forces Flight Surgeon, and even to this day, I have kept a "leadership book" that I use to take notes on the noteworthy leadership examples I have seen that I never want to forget. I used this book to study both leadership and the virtues required of warriors to succeed. The truest virtue of a warrior and the greatest leadership trait are one and the same: Love. Love of God, Love of your brother in the foxhole, Love of Country, Love of Truth. It was this love that called me to be a whistleblower and stand for those who had no one to stand for them. After I was revealed as a congressional whistleblower, other whistleblowers, including Commander Rob Green, have come forward to spearhead the fight for Love and Truth. Like my drill sergeant in 1983, Commander Rob Green seeks to overcome the current decimation of our military, rectify the wrongs done, and ultimately return our military to the heights of readiness required for our national defense. Rob carries on his back the hopes and careers of service members looking for leaders who will defend their constitutional rights and rebuild trust with the American people."

—Dr. Pete Chambers, Green Beret, flight surgeon, and
Lt. Colonel US Army (Ret.)

"When I departed home, Atlanta, for my first duty station, Fort Sill, Oklahoma, in October 1983, my Dad, US Army Corporal Herman West Sr., World War II Veteran, sent me off with a simple admonishment, and charge. He said to me, the first commissioned officer in the family, 'Son, always take care of your men.' This was a guiding principle for me in my twenty-two years of service, which included combat tours of duty. It did not mean coddling my soldiers, but training and preparing them with a sincere heart and always looking out for their well-being, and honoring my oath to the Constitution. Sadly, during the episode of COVID, there were many uniformed leaders in our military who should have heard my Dad's wise counsel. In his book *Defending the Constitution behind Enemy Lines*, US Navy Commander Robert A. Green Jr. exemplifies the courage and commitment to our sailors, soldiers, airmen, and marines that is lacking in the senior ranks of our armed forces. The COVID shot mandate for our military was an illegal, immoral, unlawful, and unconstitutional order. It was deeply rooted in politics, to the detriment of our troops. Commander Green puts it on the line in addressing this issue and the failure of military leadership that resulted in the lives of our men and women in uniform being adversely affected. There is no doubt that those who refer to themselves as senior military leaders will attempt to attack, demonize, and punish Commander Green, instead of admitting their wrongdoing in protecting our troops, and honoring their oath to the Constitution. There is a reason why recruitment and retention in our military is at disturbing lows. Commander Robert Green gives a major reason why: the betrayal of our troops. This is a must-read, and let us lock shields and protect US Navy Commander Robert A. Green Jr. from acts of retribution."

—Allen B. West, Lt. Colonel US Army (Ret.), Representative 112th US Congress; executive director, American Constitutional Rights Union

"At its core, this is a story about moral courage; the courage to do the right thing in the face of fire. This book illustrates toxic effects the cultural war waging in America has on our military, as traditional values and our constitutional republic are under assault from domestic threats. Political correctness within the military culture results in leaders at the highest levels issuing orders ethically wrong and likely illegal regarding the mandatory COVID-19 vaccine, as documented by Commander Robert Green. This is a story of failed leadership, causing distrust in our military."

—Joe Arbuckle, Major General US Army (Ret.)

"Not enough has been written or said about the loss of trust and confidence military service members have today in their leadership. It is no wonder, since today's military leaders have indoctrinated service members in the Marxist concepts of CRT/DEI and then followed up with a tyrannical, shortsighted, unneeded, harmful, and (most probably) unlawful vaccine mandate. If you think service members must blindly follow orders, you are gravely mistaken. Commander Green articulately documents the reasons why courageous young men and women believed the mandate was illegal and why they had a moral obligation to stand up against the tyranny. What certainly is not debatable is the way DOD implemented the order, which was clearly illegal given the laws surrounding the use of EUA products. I strongly urge Americans to read this book and encourage our elected leaders take action to right the many wrongs that have been committed against such incredible patriots!"

—Rod Bishop, Lt. General USAF (Ret.), chairman of the board, Stand Together Against Racism and Radicalism in the Services (STARRS)

"*Defending the Constitution behind Enemy Lines* is compelling, enlightening, and energizing. After reading the book, you'll have a deeper understanding of the issues at stake with the vaccine mandates. You'll know more about the laws that govern our land and the forces that are taking away our freedom."

—Mitzi Perdue, author, speaker, and anti-trafficking advocate

"Rob is bold and authentic, and has drawn aside the curtain for the American people allowing them to see and understand why so many military service members chose to refuse the orders of their senior military leaders during the COVID whirlwind and to stand instead for liberty. The work serves as a valuable boots-on-the-ground testimonial about the military's misplaced priorities under the control of Joe Biden and Lloyd Austin, which is a leading cause of the military's difficulty recruiting and retaining an all-volunteer force."

—Matthew Lohmeier, former Space Force Lt. Colonel and author of *Irresistible Revolution*

"If there was ever a book to be read back to front, this may be it. The good news in the concluding chapter of Commander Green's powerful saga of truth speaking to power is that the persecution suffered by the US service members who refused to comply with the Medical Dictatorship paid off in the end—if not for them, then for every American who values liberty. After reading *Defending the Constitution behind Enemy Lines*, I have never been more eager to say to these few and proud and heroic men and women in uniform: 'Thank you for your service!'"

—Michael J. Matt, *The Remnant*, editor

"Commander Green exemplifies true moral courage in writing *Defending the Constitution Behind Enemy Lines*. The illegal and immoral coercion to force the COVID-19 vaccination upon our service members was a direct violation of the law and our Constitution, the very one that those brave men and women volunteered their lives to protect, preserve, and defend. Our brave service members had their rights stripped from them before having their right to serve stolen from them, not for merely refusing the vaccine, but because they were freethinkers who came to their own logical conclusions against an unlawful politically motivated vaccine mandate. Make no mistake about it, this was a blatant attempt to remove the freethinkers from military service and it is atrociously un-American."

—Chad Robichaux, founder & CEO of Mighty Oaks Foundation, Force Recon Marine, and bestselling author of *Saving Aziz*

"In the face of silence and betrayal from Department of Defense leadership, Navy Commander Robert Green Jr. provides a riveting journey through DoD lies, deceptions, and violations of federal law. I know Rob personally from more than two years of his volunteer work with the Truth for Health Foundation. His courage, his deep faith, and his absolute commitment to his Oath to the US Constitution have given me hope that in the face of this evil assault on our core God-given freedoms, the American spirit and God's Truth will ultimately prevail. Commander Green's book is an inspiring and critical read for ALL Americans and people around the world."

—Elizabeth Lee Vliet MD, president and CEO, Truth for Health Foundation; author of *The COVID Vaccine Injury Treatment Guide: Your Roadmap to Recovery*

"To truly understand a conflict, it is always best to learn from those who survive after being completely immersed in the fight. Rob Green is one of those warrior leaders the nation can learn from as we look to right the ship of our republic. I highly recommend this read to fully understand what we are up against as a nation. This is truly a page-turner!"

—L. Todd Wood, USAFA '86, founder CDMedia

"The media portrayed the service members that refused the COVID vaccine as insubordinate, possibly ignorant extremists. The truth is the exact opposite. They were individuals who knew exactly what obedience and duty meant, understood the risks and potential moral problems with the vaccines, and worked diligently within the constitutional order of our land. In short, they were the exact kind of men and women you want fighting for your country. Rob Green calls them the 'silent minority.' This book breaks that silence with courage and clarity."

—**Jason M. Craig, executive director of Fraternus and editor of** *Sword&Spade* **magazine**

"*Defending the Constitution behind Enemy Lines* chronicles one of history's most inconspicuous fights against tyranny. It's a fight represented by the brave members of our nation's military who stood firmly against an unlawful military vaccine mandate. Some lost. Some won. But more than that, many of the country's enemies have been exposed. Be compelled to rise up against those who destroy liberty with this must-read book for the ages."

—**J. M. "Liberty Cannon" Phelps, freelance journalist**

"Commander Robert A. Green Jr. epitomizes the moral courage that should be possessed by all US military officers but is sorely lacking within the highest ranks of uniformed and civilian DoD leadership. His thorough documentation of the unlawful and immoral actions of the military establishment to coerce service members to submit to the experimental COVID 'vaccines' makes it clear that those enforcing the mandate have violated the Constitution they took an oath to protect. Green's book also makes clear that the DoD's early focus on the COVID-19 virus quickly morphed into an obsession, both with the virus and the 'vaccination status' of service members. This obsession had more to do with political agendas than the health of the force, and the resulting devastation to military readiness will have national security implications for generations to come. Despite having to fight a thoroughly politicized cadre of flag and general officers, a few good commanders like Commander Green and the leaders he highlights in the book, did everything they could to follow the law and protect the service members in their charge. Those at the top, however, consistently ignored the warnings, the scientific data, and the law to 'purge' the critically thinking leaders within the military who were still defending the Constitution. This book courageously shines the bright light of truth upon the tragic betrayal by leaders who abandoned their oaths to the Constitution. The public awareness that this book will bring is the first step in holding the perpetrators accountable."

—**Tommy Waller, president and CEO of the Center for Security Policy, Lt. Colonel USMC (Ret.), and former CO, 3d Force Recon Company**

DEFENDING THE CONSTITUTION
BEHIND ENEMY LINES

DEFENDING THE CONSTITUTION BEHIND ENEMY LINES

A STORY OF HOPE FOR THOSE WHO LOVE LIBERTY

ROBERT A. GREEN JR.

Skyhorse Publishing books may be purchased in bulk at special discounts for sales promotion, corporate gifts, fund-raising, or educational purposes. Special editions can also be created to specifications. For details, contact the Special Sales Department, Skyhorse Publishing, 307 West 36th Street, 11th Floor, New York, NY 10018 or info@skyhorsepublishing.com

Skyhorse® and Skyhorse Publishing® are registered trademarks of Skyhorse Publishing, Inc.®, a Delaware corporation.

Visit our website at www.skyhorsepublishing.com.

10 9 8 7 6 5 4 3 2 1

Library of Congress Cataloging-in-Publication Data is available on file.

Hardcover ISBN: 978-1-5107-7807-8
eBook ISBN: 978-1-5107-7808-5

Cover design by Brian Peterson

Printed in the United States of America

This book is dedicated to Our Lady of Liberty, and to all courageous patriots who risk their lives, livelihoods, and sacred honor to keep our Founding Fathers' dream alive.

Contents

Preface   *xv*

Chapter 1   Against All Enemies, Foreign and Domestic   1

Chapter 2   Sons of Liberty   5

Chapter 3   The Militarization of Safety   9

Chapter 4   Pandemic of Fear   17

Chapter 5   The Military's Response to a Near-Peer Virus   27

Chapter 6   Accommodating Equal Rights of Conscience   35

Chapter 7   Constitutional Tap Code   43

Chapter 8   The Nuremberg Shrug   53

Chapter 9   The Best Defense Is a Good Offense   61

Chapter 10 Betrayal, Exposure, and Illness   69

Chapter 11 Unconventional Lawfare   79

Chapter 12 The Test Passers   87

Chapter 13 The Court-Martial of Courage   95

Chapter 14 Thwarting the Destruction of Readiness   105

Chapter 15 Continuum of Harm   115

Chapter 16 The Cover-Up   125

Chapter 17 The Whistleblower Report   135

Chapter 18   Awakening the Sleeping Giant of Liberty   147

Chapter 19   The Call to Arms   155

Epilogue   *159*

Acknowledgments   *165*

Author's Note   *167*

Notes   *169*

Index   *187*

Preface

This story must be told. That is the conclusion I came to in 2022. By that point in time, a significant number of US military members, including myself, had been fighting to uphold their oaths to support and defend the Constitution for more than a year. We faced orders for a COVID-19 vaccine mandate that violated the Constitution, the law, and our own consciences. While some of our number succumbed to the subsequent coercion, discrimination, retaliation, and threats to our livelihoods, many of us saw those subsequent actions as confirmation of the unlawfulness clearly occurring before our eyes. Because we took our oaths seriously, this confirmation served to reinforce our resolve to continue the fight and see it to the end, though it could mean the loss of our livelihoods or even our lives.

This book will tell the story of a silenced minority who faced significant challenges with faith and courage. It is my hope that this book gives voice to the faith-filled middle-American patriots who have, until recently, made up the bulk of our armed forces. The heroes of this story are many, including tens of thousands of service members, hundreds of thousands of federal workers and contractors, and millions of citizens who helped stem the tide of tyranny in their various communities. As many heroes as this story contains, readers may be surprised that I believe the true number of villains is remarkably small.

In an ideal world we would be able to trust our government. In an ideal world a government would not seek to consolidate and wield power for power's sake. However, the ideal world rarely matches reality. The vast majority of those who went along with the unlawfulness I will lay out in this book did so because they trusted their government and the leaders appointed over them. It is likely that most of these acted in good faith and truly cared about doing the right thing. This book is not written in an attempt to prove wrong those who acted in good faith. Rather, I intend the following story to bring understanding to all who may not share my own perspective or the perspective of the tens of thousands of service members who did not go along.

In telling this story it is important to prepare the reader for two limitations that can be laid at the author's feet. First, I could never do literary justice to the sheer number of heroes who took similar actions as those described in this book. I tried to meet, communicate with, and befriend as many of these patriots as I could. However, being only one person, there was no human way I could meet and learn the stories of every patriot who chose the law and their own consciences over compliance and complicity with unlawfulness. This book only tells a tiny portion of those related stories. As in most major conflicts throughout history, the unnamed and silent heroes bore the brunt of the battle and in so doing enabled the eventual changing of the world for the better. If you are reading this and you are one such patriot whose name was not mentioned here, this book is still meant for you. This book was dedicated, in part, to you. Even if I could not tell or did not know your story, you have my deep respect and gratitude for what you have done in standing for truth and justice in the face of overwhelming coercion.

Second, I must tell this story from my own perspective, detailing events that transpired in the context of my own lived experiences and my faith in God. I attempt to draw limited conclusions about these events and in some cases provide strategic implications regarding the various actions taken. Despite the fact that my perspective cannot be understood outside the context of my faith, I make every attempt to write this book for a very broad audience. This book is written for all Americans, regardless of individual belief systems, backgrounds, or medical choices. Whether or not readers agree with the actions laid out in this book, it is my intention that all Americans will at least be able to understand our thought processes, decision-making criteria, and what we intended by our action. If this story cannot be understood by all audiences, it is likely that my own limitations as an author and a storyteller are to blame.

In living through the events I share in this book, I often found myself wondering how we came to be living through such a time as this. This particular mental exercise made it obvious to me that we, as a nation, had forgotten a number of relevant historical lessons in the years building up to what is now known as the COVID-19 pandemic. Even after nearly two years into this fight we are still overwhelmingly met by confusion from those who did not take the same approach we did. The typical questions included:

"Why can't you just follow orders?" "Why don't you just do what we did and comply?" "Don't you understand that people are dying?" Echoing in my head when faced with such questions was the aphorism that "those who forget the lessons of history are doomed to repeat it." I assert that the confusion many had regarding the stance we took in 2021 and 2022 was fueled by a lack of historical perspective required to understand our actions. Essentially, the slow progression of government overreach had finally arrived at the last possible line of defense; the bodily integrity of our very persons and those of our families.

The Constitution is the bedrock and founding document establishing the construct, the rule of law, and the civilization we have built as Americans. Those who wish to undermine the Constitution, and strip away the rights enshrined therein, have aligned themselves against a stalwart group that has taken an oath to defend the Constitution against those very actions. As military members we are used to thinking about enemies of the Constitution as being some sort of foreign threat. However, the battle lines have shifted, and domestic threats, including some from within our own government, are now coming after our inalienable Constitutional rights. These domestic threats have energized a group who takes their oaths seriously and will not stand idly by while the Constitution is dismantled around us. The following story tells the tale of courageous Americans who, when trapped behind enemy lines, elected to place their sacred honor before all else. We will never let the flame of freedom die. I wrote this story to ensure that posterity records at least some of our experiences. This story, and more importantly, the lessons imparted through it, must not be lost to future generations.

CHAPTER 1

Against All Enemies, Foreign and Domestic

Our Constitution was made only for a moral and religious people.
It is wholly inadequate to the government of any other. [1]
                                                    —John Adams

There is a cultural cold war occurring right now in the US military. In many ways this cold war is an echo of the various value-system conflicts raging throughout our nation. Since our earliest days, American values have been in flux. The cultural battle lines have shifted over time, often in response to the impact and prominence of influential individuals, organizations, and movements. In recent years, however, the erosion of traditional values has accelerated significantly. I believe the resulting incongruities would be unrecognizable to our Founding Fathers, including the normalization of alternate lifestyles, an expansion of abortion access, and a general shift toward socialism.

The military has felt the effects of this recent value-system shift as well. As an institution, the US military has historically been a bastion of traditional values. The shift away from this appears orchestrated. Military policies have been introduced that promote and institutionalize things like transgenderism and critical race theory. I have been a first-hand witness to this shift, and deeply concerned about how the pivot away from traditional values may impact the military over time. Despite my concerns, I never encountered an unlawful order related to transgenderism, critical race theory, or even white extremism during my time in the military.

As harmful as they may be, deliberately mis-prioritizing efforts, enacting purely political policies, or perpetuating bad leadership are not inherently unlawful actions. The Manual for Court-Martial states that an order is lawful "unless it is contrary to the Constitution, the laws of the United States or lawful superior orders." Normally, all orders requiring the performance of a military duty are inferred to be lawful. This means that under normal circumstances, each military order is assumed to be lawful. The Manual for Court-Martial states that these orders are "disobeyed at the peril of the subordinate." A subordinate who takes the risk of disobeying such an order is then potentially subject to a military judge for the determination of lawfulness. The Manual for Court-Martial specifies, however, that an order "must not conflict with the statutory or constitutional rights of the person receiving the order." The assumption and inference that an order is lawful does not apply to a patently illegal order that violates the law or the Constitution. As the Manual for Court-Martial notes, the inference of lawfulness "does not apply to a patently illegal order."

When the military began issuing new policies related to covering medical costs for service members seeking a gender transition, I was directed to attend a "transgenderism in the military" training session. I did not find this order to be unlawful. I was not being asked to do anything that would violate my religious beliefs. Therefore, attending the training did not violate my constitutional rights under the First Amendment. No law was broken by this training, so my statutory rights were not violated. I therefore dutifully attended the training as directed.

The Department of Defense extremism stand-down is another example of a potentially controversial order that was nonetheless lawful. On February 5, 2021, Secretary of Defense Lloyd Austin directed the Department of Defense to conduct a department-wide training about extremism in the military. As the executive officer of my maritime security squadron at the time, I was one of the leaders responsible for planning and giving this training. I reviewed the provided scripts to make sure I could, in good conscience, give this training. My review revealed that the case studies within the provided scripts focused exclusively on the threat posed by neo-Nazi extremists and white supremacists.[2] Despite later admitting that there were less than one hundred cases of extremism in the ranks,[3] the Pentagon spent over 5.3 million personnel hours planning and executing this training. The project was such a mis-prioritization and a waste of funds that a bipartisan

Senate Armed Services Committee report concluded that "spending additional time and resources to combat exceptionally rare instances of extremism in the military is an inappropriate use of taxpayer funds, and should be discontinued by the Department of Defense immediately."[4] While it may have been a massive mis-prioritization based on the number of cases found, nothing about giving this training violated my constitutional or statutory rights under law. I therefore dutifully gave the training while ensuring I pointed out that there were many types of extremism, not just what was provided in the scripted part of the training. I also encouraged my team to report any type of extremism they saw so that the chain of command could appropriately deal with it.

The determination of lawfulness should be an instinctive part of the calculus every military leader walks through in the course of receiving and disseminating orders. In an ideal world no military member should ever encounter an unlawful order. They should never receive one, and they should certainly never give one. Unfortunately, history is replete with examples of unlawful orders. Unlawful orders range the "spectrum of harm," beginning with seemingly small things like a senior officer ordering a junior officer to perform the senior's personal household chores unrelated to the junior officer's military duty. The most egregious case of unlawful orders in recent historical memory took place at the My Lai Massacre during the Vietnam War. On March 16, 1968, US Army soldiers, on orders from Army Lieutenant William Calley, murdered over 500 women, children, and old men at Sơn Mỹ village in Quảng Ngãi Province. Even the Army's response to My Lai was chilling, with an official cover-up executed by some of the highest-ranking members of Army leadership.[5] One may hope they never encounter such a thing in their career, but hope is never an acceptable plan of action. A leader who does not steel him or herself for what they will do in the event they encounter an order that violates the law or Constitution is not fit to bear the weighty responsibility that comes with that leadership.

Officers in our Armed Forces have a special role related to ensuring the lawfulness of orders. This role is highlighted by the differences between the officer and enlisted oaths of office. Both oaths include the line, "I will support and defend the Constitution against all enemies foreign and domestic, and I will bear true faith and allegiance to the same." This is the point at which the oaths begin to diverge. Enlisted members then swear to "obey the orders of the President of the United States, and the orders of the officers appointed over me." The officer oath has no such obligation. What is meant by this difference, and why would the oaths be different at all? All service members must follow lawful orders and disobey unlawful orders. However, officers have a higher obligation to ensure that the orders they receive, as well as the orders they give, are both constitutionally and statutorily sound. If an order violates the Constitution or the laws of the United States, that order loses the inference of lawfulness. In fact, numerous court cases have affirmed the obligation that service members have to resist unlawful orders. Officers are the screen through which orders are double-checked for these two critically important requirements.

There is a balance required, of course. What if every single service member paused for a constitutional analysis of each order they received? This would likely result in reduced combat effectiveness. It is also likely that good order and discipline would be degraded if every service member at every rank took the time to double-check the lawfulness of every order that was passed down. This is why the inference of lawfulness is so important. I can be lawfully ordered to take a hill in battle even if that objective does not appear to me to benefit our larger operational plan. Even a steep cost in lives to take that hill does not make the order unlawful. Service members at lower ranks or at lower echelons in the chain of command must be able to trust that those above them have taken care to ensure the lawfulness of their orders. However, if no one ever checked the lawfulness of orders, with everyone from the highest levels of the chain of command down to the lowest simply trusting that someone else was taking care of it, we would have unlawful orders slipping through the cracks with no one being held accountable. That is why it is important to balance the legal and constitutional analysis of orders with their execution.

Our forebears understood this problem and built in a natural break between the larger portion of our force (enlisted members) and the smaller portion of the force (officers) to ensure that some portion had a higher allegiance to the Constitution than to any other duty. Enlisted members have a sworn obligation to both the Constitution and to following the lawful orders given to them. Officers, on the other hand, have a sworn obligation to the Constitution only. Officers still have a duty to follow *lawful* orders, but the oath they take is to the Constitution. The Constitution is where an officer must place their primary obligation. This is done intentionally to ensure that if an officer's duty to follow an order ever conflicts with their obligation to the Constitution and the laws established by it, the Constitution and the law must take priority every single time. It is to the laws of the United States, and to the Constitution upon which those laws are founded, that officers must go to ensure the lawfulness of orders they receive and promulgate.

CHAPTER 2

Sons of Liberty

The love of liberty is interwoven in the soul of man and can never be totally extinguished. There are certain periods when human patience can no longer endure indignity and oppression.
The spark of liberty then kindles into a flame.[1]

—Samuel Adams

To best understand the modern defenders of the Constitution, it is helpful to reach back into our own history and trace the path forged by our Founding Fathers. Despite the technological differences, the conflict confronting our forefathers bears some striking similarities to the environment Americans find themselves in now. The tyranny our Founding Fathers sought to defeat began as a slow encroachment into the lives and rights of the colonial people of the 1700s. The resulting American Revolution was not a hurried response to an instantaneous infringement of colonial rights, but was a slow awakening of a people wearied by long oppression and subjugation. The events that ultimately ignited the American Revolution were not so much a flashing heat, but slowly smoldering coals fanned into flames by the increasingly tyrannical actions committed by an overbearing governmental power.

The story surrounding the transformation of the colonial peoples from loyal British subjects to revolutionists can hardly be told without mentioning Samuel Adams. Although the term whistleblower was not coined until nearly two centuries later, Samuel Adams played a critical role in exposing to public scrutiny the violence and tyranny being inflicted on colonial subjects. He wrote extensively in the years leading up to the American

Revolution and carefully crafted his message to rouse his fellow country-men to join him in resisting tyranny. He even wrote his Harvard master's thesis on the lawfulness of resisting British rule, asserting that it was lawful to resist if there was no other way to preserve the commonwealth.[2] Samuel Adams's role in winning over his countrymen was so critical that many biographers argue that it is unlikely the American Revolution would have happened without him.[3]

While it is typical for some injustices and resentments to occur between governments and some of the peoples they govern, the relationship between the British Empire and the general colonial public began to sour in a broader way following the French and Indian War. In an effort to quell future possi-ble conflicts with Native American tribes, King George III issued the Royal Proclamation of 1763 which forbade new settlements inland, prohibited pri-vate land purchases from natives, and dictated that only licensed traders could trade with Indian tribes. While some colonists understood the king's intentions, many completely ignored the proclamation.[4] This was followed by both the Currency Act and the Sugar Act in 1764. The Currency Act prohibited the colonists from issuing paper currency. The Sugar Act placed a tax on sugar and molasses imports in an effort to gather funds to pay war debt incurred during the French and Indian War.

While most colonists had little reaction to the Sugar Act, Samuel Adams jumped at the chance to demonstrate the infringement of rights implemented by parliament through a tax on a people who had no represen-tation in the body enacting the tax. Adams began shifting colonial percep-tions by making a point to warn the Massachusetts Assembly in 1764 that the Sugar Act was likely "preparatory to new taxations" upon the colonists.[5]

The Sugar Act was just the beginning. It was followed by the Stamp Act in 1765 which taxed paper used for printed material, and the Townshend Acts which, amongst other things, added heavy taxes to consumer goods and targeted smugglers by forcing them to be tried in admiralty courts without juries. The Stamp Act incited the first public demonstrations and protest by the colonists including the August 1765 Stamp Act Riots in Boston. Samuel Adams and his growing organization, the Sons of Liberty, led boycotts of British goods in response to the 1767 Townshend Acts.

The British colonial government sent four thousand troops to Boston in 1768 which further escalated tensions and cemented distrust in the colonial population of Boston. The Tea Act of 1773 provoked the Boston Tea Party in December of that year. Rather than back down and attempt a new approach, the British Parliament's response to the Boston Tea Party tipped the scales towards revolution by enacting the Coercive Acts of 1774 (also called the Intolerable Acts). The Intolerable Acts closed the port of Boston, restricted democratic town meetings, made British officials immune to criminal prosecution in Massachusetts, and required colonists to house and quarter British troops on demand in their private homes.[6]

The relatively rapid succession of subsequent tax and control initiatives targeted to increase revenue and quell dissent in the American colonies had ancillary consequences. As the British Empire became more aggressive in taxing and attempting to control the colonial population, more colonists were driven into Samuel Adams's ideological camp. The Sons of Liberty swelled with new members, including some of the most highly educated and influential Americans of that time. Membership in the Sons of Liberty included John Hancock, Patrick Henry, Benjamin Rush, Paul Revere, and even Benedict Arnold. The Sons of Liberty played an important role in ensuring that the Intolerable Acts did not accomplish what the British intended.

Instead of suppressing the growing resistance, the British response in Massachusetts was seen as so tyrannical that other colonies were convinced to come to their aid. The tyrannical overreach committed by the British government had finally reached the last possible line of defense for many Americans, their very homes. It was necessary and inevitable that the colonists would defend their homes and their way of life against the Quartering Act and other tyrannical British actions. The Declaration of Independence and the American Revolution were the pivotal and decisive responses.

Following the American Revolution and the consummation of the split from the British Empire, our Founding Fathers, a great many of whom had been Sons of Liberty, were left trying to figure out what sort of government to put in place. The debate over the Constitution by the Federalists and the Anti-Federalists was a heated one. The Federalists wanted a strong centralized government and felt that the Constitution did enough on its own to protect individual rights, rendering a separate Bill of Rights unnecessary. The Anti-Federalists did not want to ratify the Constitution at all, fearing that a strong centralized government, like the one they had just overthrown, would fail to protect individual rights. A compromise was eventually reached through a ratification of the first ten amendments that enumerated specific rights, commonly known as the Bill of Rights.[7]

Individual medical freedom was not one of the rights specifically enumerated by the Bill of Rights. Medical freedom is still protected by the Constitution, however, through the Fifth, Ninth, and Fourteenth

Amendments in particular. Despite these protections, the concerns that motivated the Anti-Federalists nearly 240 years ago have proven prophetic. The ink had barely dried on the Bill of Rights when the federal government began the slow progression of consolidating power and restricting the liberties of individuals. In the last fifty years or so, the federal government has only accelerated the consolidation of power and the encroachment on individual liberties, particularly in the area of medical freedom. The current individual medical freedom crisis is the inevitable conclusion of a governmental power attempting to control and standardize public health at the national level without regard to individual medical freedoms protected by the Bill of Rights. How this crisis plays out is up to us and our willingness to resist like the Sons of Liberty, who played such a crucial role in birthing the Constitution we now swear to defend.

CHAPTER 3

The Militarization of Safety

Those who would give up essential Liberty, to purchase a little temporary Safety, deserve neither Liberty nor Safety.[1]
—Benjamin Franklin

The absence of specifically enumerated medical freedom rights within the Bill of Rights should not be taken as an indication that our Founding Fathers would not have cared about medical freedom. I assert that our Founding Fathers would find the premise behind our current medical freedom crisis repugnant. They lived in a time when medicine was rudimentary, and Americans were dying of diseases, natural disasters, and general accidents at rates that would be alarming to modern sensibilities. Safety from these perils, however, is not what colonial Americans asked for from the British. In a 1775 speech urging fellow Virginia Convention members to vote for an independent militia, Patrick Henry did not conclude with "Give me *safety*, or give me death." That would have been paradoxical, and no one would have voted for that. Rather, Patrick Henry concluded his speech with a fiery "Give me liberty, or give me death!"

The Third Amendment is informative in this situation. The Third Amendment states that "no soldier shall, in time of peace be quartered in any house, without the consent of the owner, nor in time of war, but in a manner to be prescribed by law." The Third Amendment was our Founding Fathers' response to the Quartering Act of 1774, which was the last of the Intolerable Acts, and the only one which applied to all the colonies.[2] The Intolerable Acts, and the British attempt to enforce them, was the final step that ultimately pushed the American colonies too far and incited a collective response to resist. Upon winning the war and establishing their own government, our Founding Fathers elected to specifically include the rights protected by the Third Amendment. As the Ninth Amendment states, this does not mean that other rights, including individual medical freedom rights, did not exist. Our Founding Fathers just happened to face the invasion of their properties and homes by the British, who wrongfully asserted a right to do so through the Quartering Act. Had the British been forcing unwanted or unnecessary medical treatments upon colonial Americans, I am confident that our Founding Fathers would have enumerated a specific individual medical freedom right within the Bill of Rights in response. The Constitution, despite not specifically enumerating the right, does protect individual medical freedoms. Various legislatures and even Supreme Court rulings, however, have wrongfully chipped away at these rights over the past two centuries.

Although the Constitution is not a perfect document, it has stood the test of time for nearly 240 years. It has endured assaults from external attacks and from internal pressures. Our Founding Fathers attempted to build broad protections into the Constitution to ensure it could survive all future assaults, not just the assaults they could explicitly foresee. Unfortunately, the assaults on our Constitution stemming from internal pressures are self-inflicted wounds. Every bad law devised by man, voted in by a legislature, and subsequently deemed unconstitutional by the Supreme Court, is a successfully thwarted assault on our God-given rights and liberties that the Constitution protects. Our Founding Fathers understood these rights to be inalienable because they were founded in higher moral law. The Constitution, as a proxy for this higher moral law, was designed humbly enough to account for changes and moral evaluations of new laws to determine whether those new laws conformed with natural law. Since its ratification, the Constitution has served this role well, with the Supreme Court ruling a total of 644 laws as unconstitutional since our founding.[3] Each one of these constitutional assaults, successfully parried, was initiated by misguided or nefarious humans at the helm of various legislatures.

Of greater concern than legislators writing unconstitutional laws, are Supreme Court rulings that incorrectly interpret the Constitution. Rulings of this type enable the continued enforcement of laws that violate the Constitution and potentially override rights that would otherwise have been protected by the Constitution. Since the founding of our nation, there have been 234 Supreme Court rulings which have been subsequently overturned by later Supreme Court rulings;[4] forty-five of these overturned rulings were in place for longer than fifty years. One Supreme Court ruling was 136 years old before being overturned by a later ruling. During the intervening years before the Supreme Court could correct a prior ruling, how many unconstitutional laws were written and enforced under the presumption of constitutionality? It is possible that an incorrectly decided Supreme Court ruling can enable years of unconstitutional laws and an expansion of powers not permitted by the Constitution before a later Supreme Court ruling can correct the mistake. It is also not a stretch to assume that the longer a ruling goes before being overruled, the worse its impact on our nation and the rights that should have been protected under the Constitution.

It is within this context that we will review the Supreme Court rulings that have had the biggest impact on the current individual medical freedom crisis and the COVID-19 vaccination mandates being imposed on American citizens. The 1905 *Jacobson v. Massachusetts* Supreme Court case upheld state authority to impose immunization requirements and has been called "one of the most important pieces of public health jurisprudence" by some legal scholars.[5] In an effort to provide safety to their community from a local outbreak of smallpox, Cambridge, Massachusetts enacted a 1902 law that required all residents to be vaccinated. Anyone over twenty-one years of age who refused to be vaccinated would be fined a one-time $5 penalty (a little more than $150 today). There was no provision to vaccinate any citizen by force. Local Lutheran Evangelical Minister Henning Jacobson refused the smallpox vaccine, citing a bad reaction from a prior vaccination that left him needing years of special care for a burning rash.[6] At both his trial and on appeal, Jacobson did not raise any First Amendment arguments, nor did he cite his religious convictions in his defense.[7] Jacobson was convicted at trial court and lost on appeal to the Massachusetts Supreme Judicial Court.

Of note, the Massachusetts Supreme Judicial Court determined that any harm resulting from vaccination was not pertinent to the case due to the fact that *the law did not permit the state to actually vaccinate Jacobson against his will.*[8] In the unanimous majority opinion, Massachusetts Chief Justice Knowlton wrote, "If a person should deem it important that vaccination should not be performed in his case, and the authorities should think otherwise, it is not in their power to vaccinate him by force, and the worst that could happen to him under the statute would be the payment of the penalty of $5."[9] In essence, the court acknowledged that the compelling government interest in providing safety and security in the presence of an outbreak of disease did not grant the power to compel vaccination.

Henning Jacobson appealed this decision to the United States Supreme Court which handed down a 7–2 ruling against him and upheld the $5 fine levied by the Cambridge Massachusetts Board of Health.[10] The ruling confirmed that individual liberty is not absolute and can be restrained "in order to secure the general comfort, health, and prosperity of the State."[11] Despite ruling in favor of the state, the Court laid out several limitations that should protect individual rights. In a work published in the *Buffalo Law Review* in January 2022, Dr. Josh Blackman detailed these limitations. First, the *Jacobson* ruling recognized that the law "cannot be enforced against a person for whom the vaccine would be particularly dangerous."[12] Second, the vaccine mandate must have a substantial relation to the public health motive it was enacted for and cannot be enacted for some ulterior motive.[13] Third, the Court acknowledged individual rights could not be violated arbitrarily or in an unreasonable manner.[14] Finally, the $5 fine was modest and on the far low end for fines during that era. Many fines, such as those for integrating classrooms or for corrupting public water sources were orders of magnitude higher.[15]

The case was essentially about the imposition of a financial penalty on an individual's right to refuse both the fine and the state requirement the fine was meant to enforce. Of note, *Jacobson v. Massachusetts* did not rule on or even review the competing interests between the state's desire to provide safety and an individual's First Amendment religious freedom rights. As Dr. Blackman correctly points out, *Jacobson* "did not, and indeed could not, resolve the question of whether the state could force a person to undergo a medical procedure."[16] *Jacobson v. Massachusetts* was a *narrow* case with a *narrow* ruling.[17] In his conclusion, Justice Harlan even specified that the *Jacobson* ruling decided "only that the statute covers the present case," confirming the Court's intention that the ruling be interpreted narrowly.[18] Despite the limited scope of *Jacobson v. Massachusetts*, later Supreme Court rulings have greatly—and inappropriately—expanded the meaning and application of what was once a very narrow precedent.

The most significant Supreme Court ruling that reshaped and expanded the meaning of *Jacobson v. Massachusetts* was also one of the most notorious in the Court's history. The 1927 *Buck v. Bell* Supreme Court ruling was a human rights travesty that is, unbelievably, still in place today despite the Court having multiple opportunities to overturn it.[19] *Buck v. Bell* upheld a 1924 Virginia law permitting the state to sterilize the "feebleminded" for the welfare of society, and thereby eliminate the propagation of "socially inadequate offspring."[20] This ruling came at a time in our nation's history when eugenic Darwinian social engineering was all the rage and society was deeply concerned about the propagation of undesirable segments of the population.[21] Justice Oliver Wendell Holmes, who wrote the majority opinion in *Buck v. Bell*, was the only remaining member of the 1905 *Jacobson* Court.[22] He was also a lifelong proponent of eugenics and one who advocated for "restricting propagation by the undesirables and putting to death infants that didn't pass the examination."[23] Justice Oliver Wendell Holmes knew very well how narrow the *Jacobson* decision was, but intentionally misread *Jacobson* in order to justify forced sterilization in *Buck*. His doing so ripped open a hole in the meaning of *Jacobson* and vastly expanded the powers of the state to provide "safety" in the face of a contrived societal danger. Citing *Jacobson v. Massachusetts*, Justice Holmes concluded that "The principle that sustains compulsory vaccination is broad enough to cover cutting the Fallopian tubes. Three generations of imbeciles are enough."[24] This ruling, horrific both in its inhumanity and in the forced sterilization that resulted, is a travesty that continues to haunt the Supreme Court and continues to undermine trust in it as an institution.

The problem with Justice Holmes's conclusion, in addition to its fundamental inhumanity, is that *Jacobson* did *not* permit compulsory vaccination and Justice Holmes cited only the *Jacobson* ruling in making the claim that there was an established principle supporting compulsory vaccination. In fact, Holmes did not cite a single other Supreme Court ruling or precedent at any point in his *Buck v. Bell* decision.[25] Holmes, essentially using *Jacobson* as a pretext, *created* a principle of compulsory vaccination that did not exist in *Jacobson*, and then used that principle as justification to create a principle supporting forced sterilization. Holmes also uses a false equivalence to justify his decision. As Dr. Blackman notes, "a single dose of a well-established vaccine cannot be plausibly compared to the permanent destruction of reproductive organs."[26] Holmes's creation of a false precedent and his use of a false equivalence were not the only significant falsehoods present in *Buck v. Bell*. The premise it was built upon, that the sterilization of Carrie Buck would provide safety from future undesirable offspring, was itself built upon criminally dishonorable lies.

Carrie Buck was born in Charlottesville, VA, on July 2, 1906, to Frank and Emma Buck. The early loss of Frank left Carrie's mother to scratch out a living on her own. She was not very successful, and a local middle-class family, the Dobbses, stepped in to offer a foster home to Carrie.[27] John and

Alice Dobbs took Carrie into their home to be more of a housekeeper than a foster child. They pulled her out of school after the fifth grade and eventually hired her out to work for neighbors in order to increase their income.[28] Carrie's life changed dramatically when she was raped by the Dobbses' nephew, Clarence Garland, when he came to visit in the summer of 1923. The pregnancy that resulted would have been a significant embarrassment for the Dobbs family, particularly if Carrie went public about how she came to be pregnant. With the nation in a panic over the supposed threat to the gene pool posed by the so-called feebleminded, the Dobbs family found a ready excuse to quickly and permanently send her away.[29] John Dobbs petitioned the Charlottesville Juvenile and Domestic Relations Court to have Carrie declared feebleminded and asked that she be committed to the Virginia Colony for Epileptics and Feebleminded.[30]

After Carrie was institutionalized, she came into contact with individuals who had been looking for a legal test-case to get forced sterilization in front of the Courts. They found their ideal victim in Carrie.[31] She had no one to advocate for her, and no one even explained to her that the operation they intended would preclude her from having children in the future. No one in the legal system, from the Charlottesville Juvenile and Domestic Relations Court to the United States Supreme Court, even bothered to verify Carrie's "feeblemindedness." Her true intelligence was either ignored or hidden from them, including the fact that her grades, until she was pulled out of school, were perfectly normal. Even Carrie's daughter, conceived in rape and who died of a stomach infection during the height of the Great Depression, was described as "very bright" and had perfectly normal grades in school.[32] None of the evidence of Carrie Buck's quiet intelligence came to light until years after the *Buck v. Bell* decision had done its damage.

Both of the last two chilling points made by Justice Oliver Wendell Holmes in the *Buck v. Bell* ruling—that "The principle that compulsory vaccination is broad enough to cover cutting the Fallopian tubes" and that "Three generations of imbeciles are enough"—are falsehoods that demand the justice of correction. Ultimately, American governments authorized between 60,000 to 70,000 forced sterilizations.[33] Most of these were enabled by *Buck v. Bell*, and by the falsehoods written in that ruling by an aging and entrenched proponent of American eugenics, Supreme Court Justice Oliver Wendell Holmes.

From a safety versus liberty perspective, *Buck v. Bell* was a monumental ruling for those who wanted to expand the power of the state. In *Buck*, Justice Holmes didn't just recast *Jacobson* as permitting forced vaccination.

He used *Buck* to reshape the size, type, and scope of the societal threat considered in *Jacobson*. In *Jacobson* the threat was from a localized outbreak of smallpox. In *Buck*, the contrived threat posed by the propagation of undesirable human beings that would supposedly decrease the welfare of society was sensationalized and then nationalized. Since *Buck* opened the door to this level of pandemic jurisprudence without regard to substantive and enumerated constitutional rights, including First Amendment religious freedom rights, the powers of the executive and legislative branches of government have vastly expanded under the auspices of safety threats and health threats, contrived or otherwise. Courts have relied on *Jacobson* for pandemic jurisprudence to make rulings about "religious freedom, abortion, gun rights, voting rights, the right to travel, and many other contexts."[34] All of this is well beyond the original scope of *Jacobson* and was enabled by Justice Oliver Wendell Holmes in the human rights travesty that was *Buck v. Bell*.

The correction to this inappropriate balancing of safety and liberty, built on false premises, is offered by Dr. Josh Blackman. He states that the "[Supreme] Court should explain that its modern substantive due process precedents govern disputes about bodily autonomy not *Jacobson* . . . [and] to dispel any doubts, the Court should limit *Jacobson* to its facts: the government can impose a modest fine on the unvaccinated, with exemptions based on specific health concerns. And, in light of modern doctrine, exemptions must also be granted to protect the freedom of religion."[35]

The *Buck v. Bell* ruling was, and continues to be, antithetical to the Constitution, to basic human decency, and to the unchanging moral law understood by people of goodwill throughout history. When people point to *Jacobson* as the constitutional precedent for the current mandates, with subsequent coercive tactics to enforce COVID-19 vaccination, they are doing so upon the shoulders of the *Buck v. Bell* ruling. Those perpetrating the COVID-19 vaccine mandates on Americans are not the first to use the precedent established by *Buck v. Bell* to justify their actions. At the 1945 Nuremberg trials, Otto Hofmann, the head of the SS Race and Settlement Office, quoted Justice Oliver Wendell Holmes's ruling in *Buck v. Bell* as justification for the mass sterilizations he perpetrated against "obviously inferior individuals" on behalf of the Nazi party.[36] This defense did not work at the Nuremberg Trials and will not justify those who are unlawfully enforcing the COVID-19 vaccine mandate today. A governmental power making an individual's medical decisions, vaccinating unwilling citizens, or forcibly sterilizing women, would have been unconscionable to our Founding Fathers. Yet under the color of law, and with Supreme Court rulings that wrongfully and tacitly permit the expansion of governmental powers in these areas, modern Americans are now subject to the intolerable.

We are now living through what our Anti-Federalist forefathers foresaw and dreaded. They had to fight just to have a specific Bill of Rights included in the Constitution. Their efforts, although commendable, have proven to be inadequate to quell the expansion of federal power and government over-reach. The Anti-Federalists feared a time when a desire for safety was used as the means for trampling liberty. They did not anticipate that the safe-ty-versus-liberty battle could potentially be lost in the arena of individual medical freedom. They would have fought to strengthen the Constitution in that area had they known. Like the Sons of Liberty who faced the forced quartering of British soldiers, we have been backed down to our last possible line of defense. The defense of our own bodily integrity is more than a pro-verbial line in the sand. There is no possible further retreat. There is nowhere left to go. Those of us who have taken an oath to support and defend the Constitution have an obligation to take a stand.

CHAPTER 4

Pandemic of Fear

I know that the practice of predicting danger and death upon every occasion is sometimes made use of by physicians, in order to enhance the credit of their prescriptions if their patients recover, and to secure a retreat from blame if they should die.[1]

—Dr. Benjamin Rush

Fear can be a great motivator. Many times, having a healthy fear of something can keep a person alive. Service members on the battlefield can attest to this fact, as can anyone whose fight-or-flight reaction to a threat kept them from surrendering in the face of something that might otherwise have taken their life. In the case of sickness or death, fear can be the motivator that forces us to maintain discipline with good hygienic practices. Fear can also be taken too far, however. Unhealthy amounts of fear can debilitate individuals or drive whole communities into hysteria and panic. While fear can sometimes incapacitate the unprepared, it can also cause wild overreactions, where the fearful leap from one extreme to another as they desperately seek solutions. The COVID-19 virus, and the global reaction to it, demonstrated how fear could cause this level of desperation on a global scale. We are still reeling from the various overreactions to the COVID-19 pandemic committed by governments, global leaders, and corporations. Many of these overreactions were exacerbated by media entities, presumably in search of higher ratings. Some of these governments, organizations, and entities seemed legitimate in their desperate search for solutions; others apparently schemed to take advantage of the fear they stoked in the populace.

I remember the fear I felt during that time. It was in early March of 2020, and I remember pacing back and forth, running scenarios through my head while watching a presidential press conference about the pandemic. I had a great deal of concern for my family, but I also had a lot of questions. I did not have access to all the required data to answer my own questions about the COVID-19 virus and its impact on patients of different ages, backgrounds, and various medical conditions. I struggled with not being able to access this data and work toward finding answers to my questions. Along with millions of other Americans, I waited for whatever information the government chose to reveal or that the media could get their hands on and spread, uncensored, to the world, or so I thought. The various news broadcasts and even the government press conferences did not help me make sense of the situation. The broadcasters always seemed to focus on the most severe and serious aspects of the situation. We were told death counts, case rates, and even hospital bed shortage rates. All of this information and the way it was presented seemed to be intentionally designed to maximize a fear response. Lockdowns, masks, social distancing guidance, travel restrictions, and a full stop to our otherwise healthy economy followed.

At some point later in 2020, under the continued barrage of fear-inducing information and the never-ending mitigation measures, I grew skeptical of what I was being told. I was not skeptical about the deadliness of the virus for at-risk and elderly populations, nor was I skeptical about the need for positive action to protect the vulnerable. Rather, my growing skepticism was about the motivations and intentions of the individuals and organizations pushing for the most extreme, inflexible, and standardized mitigation measures, regardless of individual risk. There would be glimmers of hope as we learned more about the virus and its impact. Young children and healthy young adults faced minimal risk from the virus. The case fatality rate declined as doctors learned more about treatments. However, each positive development was quickly brushed aside as media entities continued to keep the public focused on whatever was more attention-grabbing. When the death rate slowed, for example, media entities shifted to keeping cumulative death counts in the corner of the news broadcast screens, ostensibly because they thought it would be more alarming.

My pro-life beliefs played a key role in finally confirming that my doubts and skepticisms were well founded. It came as no surprise to me that COVID-19 mitigation measures were becoming a very polarizing political issue. What was surprising was that politicians with near perfect pro-choice voting records seemed to be the most strident in calling for extreme mitigation measures and attempting to remove individual choice from the American people. As if they were all reading off the same cue cards, many of them used the same refrain as justification for their approach. I kept hearing that these measures were worth it "if it saved even one life." As a pro-life Christian I could not help but think of lives lost in the womb. Why were these lives not worth the same strident approach from our political leaders? What if these same politicians voted for and funded free and voluntary adoption services that respected both the life of the mother and the child? What if such measures could "save even one life"? There was an obvious double standard being applied to saving certain lives over others. The lack of consistent logic helped me realize that the motivations and intentions of governmental agencies, political leaders, media entities, and corporations were seriously suspect. This realization also helped me stop fearing, start researching, and begin critically thinking about the solutions being proposed.

The solution to the COVID-19 pandemic at the forefront of the national conversation was vaccination. We were told that a COVID-19 vaccine would put a stop to the pandemic and help get America "back to normal." According to the CDC published timeline, Operation Warp Speed was launched by the Trump Administration on April 20, 2020. This interagency partnership between the Department of Health and Human Services (HHS) and the Department of Defense (DoD), was designed to produce a COVID-19 vaccine as quickly as possible.[2] The fact that Operation Warp Speed altered the standards for research processes as well as the normal vaccine development timeline was no secret. According to a July 1, 2020 National Institute of Allergy and Infectious Diseases (NIAID) publication on the National Institute of Health (NIH) website, "Operation Warp Speed will manufacture promising vaccine candidates at an industrial scale before efficacy and safety are confirmed through Phase 3 trials. Doing so will significantly shorten the timeline for distribution as compared to traditional vaccine development, should the trials succeed."[3]

Since COVID-19 vaccine mass production was to begin *before* safety and effectiveness data were confirmed, what would be done with all of the mass-produced vaccines if the results indicated that they were unsafe? Either no one was asking this question, or I was unable to find anyone asking it because of censorship. As a non-expert in the field, and an admitted amateur, I doubted I was the only one who had this question. If the experts and professionals in the field were asking the question, however, was censorship being implemented to ensure I could not find those questions or answers?

Pondering this conundrum was the start of unveiling an agenda so contrary to the Constitution and to the public good that I had a great deal of trepidation in continuing to dig deeper. I resolved to continue to dig into the safety and efficacy of the vaccines as they became available.

Whatever history ultimately concludes about the safety or efficacy of the vaccines developed to combat COVID-19, no one can argue against the fact that the vaccines were produced at a speed unheard of when compared to all previous vaccine products. On December 11, 2020, less than nine months after the launch of Operation Warp Speed, the FDA issued the first Emergency Use Authorization for the Pfizer-BioNTech COVID-19 vaccine for all people ages sixteen and up.

The development and approval process for a traditional vaccine can be as long as twelve years from preclinical development to final review.[4] Even the NIH admits that "Many vaccines take ten to fifteen years to reach the public."[5] They claim, however, that the COVID-19 vaccine development timeline was a special case because for many years before the pandemic "experts at the NIH Vaccine Research Center (VRC) were studying coronaviruses to find out how to protect against them. The scientists chose to focus on one "prototype" coronavirus and create a vaccine for it."[6] During that same period, from 2014–2019, the NIH was also indirectly funding coronavirus "gain of function" research (research into increasing coronavirus transmissibility and lethality) through a $3.1 million grant awarded to EcoHealth Alliance.[7] Of interest to all service members and veterans, the Department of Defense also funded EcoHealth Alliance to the tune of $41.91 million since fiscal year 2008; $33.85 million of which was awarded for a work program labeled "Scientific Research—Combating Weapons of Mass Destruction."[8] We may never know exactly how much DoD money was spent by EcoHealth Alliance at the Wuhan Institute of Virology for weapons research or how exactly the COVID-19 virus originated.

The possibility exists that the SARS-CoV-2 virus was the result of either a lab leak or an intentional release from the Wuhan Institute of Virology, located within the same community that the COVID-19 virus was first detected.[9] One of the most significant studies conducted in an attempt to disprove the theory that SARS-CoV-2 originated at the Wuhan Institute of Virology included in its conclusion "the possibility [that] a laboratory accident cannot be entirely dismissed" and that a laboratory accident would be "near impossible to falsify."[10] Of significant note, and a possible indication of a conflict of interest, is the fact that multiple authors of this study acknowledged funding by the Bill and Melinda Gates Foundation and the NIH, major proponents of the subsequent COVID-19 vaccination campaigns.[11]

The origins of the COVID-19 virus have been disputed since the earliest days of the pandemic. What is not in dispute, however, is that some government funding did get funneled to the Wuhan Institute of Virology through EcoHealth Alliance. One US Army veteran, Dr. Andrew Huff, who was a senior EcoHealth Alliance vice president from 2014 to 2016, has since come forward as a whistleblower with evidence that EcoHealth Alliance was funding "gain of function" research at the Wuhan Institute of Virology. In a recent interview, Dr. Huff revealed that EcoHealth Alliance eschewed their responsibility for biosecurity and biosafety. Dr. Huff stated that "The worst part is they defer their responsibility of what's called a biosecurity officer or institutional biosafety committee to a subcontractor. That subcontractor is the Wuhan Institute of Virology."[12] In response to congressional inquiries and public revelations that EcoHealth Alliance was funding "gain of function" research at the Wuhan Institute of Virology, the NIH sent an October 20, 2021 letter to Congressman James Comer admitting that EcoHealth Alliance failed to operate according to the terms of their grant. In that same letter the NIH also informed Congressman Comer that they were putting EcoHealth Alliance on notice to provide all "unpublished data from the experiments and work conducted under this award."[13]

Despite the lack of oversight, the outsourcing of biosecurity and biosafety to China, and the significant unanswered questions about how the pandemic began, the NIH actually issued a new grant to EcoHealth Alliance on September 21, 2022 for additional coronavirus research.[14] The apparent duplicity by this segment of our government is so significant that it is hard to wrap one's mind around it without stating the situation plainly: The NIH developed a vaccine in preparation for a pandemic that they may have helped create either through negligent oversight or possibly some level of criminal complicity.[15] The study of this apparent duplicity and the questions this situation raises is beyond the scope of this work. However, the facts, as laid out here, should be enough to lead every single American to a healthy skepticism regarding the motivations and intentions of those who may benefit in some way by promulgating the vaccine.

Regardless of how the vaccine was made and the intentions of those who funded its production, there are still developmental steps that must be followed to ensure the safety and effectiveness of the vaccine. After a vaccine candidate is developed, preclinical trials are initiated to test the vaccine in animals. If the preclinical trials show promise, human testing, called clinical development, can begin. The clinical development is done in three phases. Phase 1 emphasizes safety, and the product is typically only tested on twenty to one hundred volunteers who have not been exposed to the disease yet. If there are no manifest safety concerns in Phase 1, Phase 2 is initiated to study varying dosages in as many as several hundred volunteers across different demographic groups in randomized-controlled tests. Phase 2 trials typically include a control group which receives a placebo so that the individuals receiving the vaccine candidate can be compared to the control group. In Phase 3 trials, the vaccine is generally administered to thousands of people and generates "critical information on effectiveness and additional important safety data."[16] These phases typically take anywhere from one to three years each.

To further explain the speed at which the COVID-19 vaccines were developed, the CDC, on their website, noted that in the case of the COVID-19 vaccines the clinical trial timeline was reduced without skipping any clinical trial phases because those phases were overlapped. The language describing the overlap of phases has since been removed from the CDC website. The CDC website now simply states that "No trial phases have been skipped."[17] Though the CDC appears to be trying to spin the clinical trial language on their website, the information about overlapping COVID-19 clinical development phases remains on the ClinicalTrials.gov website. Federal law requires the Department of Health and Human Services, through the NIH, to "establish a registry of clinical trial information for both federally and privately funded trials conducted under investigational new drug applications to test the effectiveness of experimental drugs for serious or life-threatening diseases or conditions."[18] The ClinicalTrial.gov website lists the Pfizer BioNTech COVID-19 vaccine trial as a "Phase 1/2/3" trial. This trial, with identification number NCT04368728, was begun on April 29, 2020 with an estimated completion date of March 15, 2023.[19]

Someone has been manipulating the end date of this trial, and rolled back the end date to January 31, 2023. Regardless of end date, those in charge of the COVID-19 vaccine products were already shipping them around the world and encouraging their administration to fear-weakened populations. The lack of a completed Phase 3 trial, which is supposed to provide critical effectiveness data along with additional information on safety, has not stopped Pfizer, our government, or medical providers from administering this product on an industrial scale. As of October 17, 2022, 372,455,530 doses of the original Pfizer BioNTech vaccine had been administered in the

United States. All of these doses were administered *before* clinical trials were completed.[20]

The fact that there were clinical trials at all is almost meaningless if you understand that the pharmaceutical companies fund and conduct almost all of their own trials. As stated in an article by Dr. Sameer Chopra, "Most clinical trials, however, are funded by pharmaceutical companies with enormous financial stakes in the products being evaluated. Furthermore, the scientists who design, conduct, analyze, and report clinical trials often receive monetary compensation from drug companies, in the form of either salaries or consulting fees."[21] With little to no oversight, these companies have a long track record of hiding trial results, throwing out negative data, and only presenting the most favorable reports for review.

In his book *Bad Pharma: How Drug Companies Mislead Doctors and Harm Patients*, Dr. Ben Goldacre describes the various unethical practices companies employ to produce favorable results during clinical trials of medical products in development. He highlights a study covering one year of randomized controlled trials in which forty-five out of forty-five trials, a perfect 100 percent, demonstrated favorable results to the sponsor of the drug under trial. The funding source was the decisive factor in every trial. As Dr. Goldacre put it, "if it was funded by industry, you could know with absolute certainty that the trial found the drug was great."[22]

The COVID-19 vaccine trials were no exception to this unethical and potentially fraudulent practice but with the added complexity that the sponsor company, Pfizer, desired speedy results in order to claim the title of "first successful COVID-19 vaccine." Brook Jackson, an experienced vaccine researcher and then-regional director of the clinical research company Ventavia Research Group, was one of the scientists conducting the Pfizer BioNTech Phase 3 trial (NCT04368728). Her company was contracted by Pfizer to provide vaccine test sites for the Phase 3 trials and was pushed by Pfizer to enroll as many patients as possible in the vaccine trial as quickly as possible. She was responsible for general oversight and quality assurance at two separate company locations. In the process of her work on the trial, she became concerned that the speed of the trial was putting data integrity and patient safety at risk.

Beginning on September 8, 2020, Ms. Jackson began reporting to her company leadership the many clinical trial protocol and FDA regulatory violations she observed. The list of violations she documented are extensive, but they include falsification of data, failure to preserve clinical trial "blinding," vaccine dilution errors, patient safety issues, use of unqualified staff as vaccinators, and failure to monitor patients post-injection. None of her concerns or any of the identified violations were addressed, prompting Ms. Jackson to make a report to the FDA. Ventavia fired Ms. Jackson later on the same day she made the report to the FDA, under the pretext that she was "not a good fit" despite the fact that she had no prior disciplinary issues, nor had she received any negative feedback regarding her job performance.[23]

The fact that a Pfizer clinical trial is once again the subject of significant questions of fraud should not be surprising to anyone paying attention to the pharmaceutical industry. Pfizer's recent past is fraught with serious ethical failures and systemic criminal conduct. According to a 2010 HealthCare Policy article by Robert G. Evans, "Pfizer has been a habitual offender, persistently engaging in illegal and corrupt marketing practices, bribing physicians and suppressing adverse trial results."[24] In an eight-year span from 2002 to 2010, the company and its subsidiaries were assessed $3 billion in criminal convictions, civil penalties, and jury awards. This $3 billion penalty may seem like a great deal of money. However, it was actually less than half of their 2009 revenue alone, making the penalty closer to a slap on the wrist than an action intended to effect real change. Individual executives were also not held accountable, leaving the corporations themselves liable for the subsequent slap-on-the-wrist financial penalties. The irony, as Evans points out, is that "In the absence of such personal liability, both criminal and civil penalties appear to be, to Pfizer at least, a business expense worth incurring."[25]

This report from 2010 has not impacted or influenced enforcement in any meaningful way. We are in the current predicament with Pfizer and other pharmaceutical companies, in large part, because we are not holding the companies and their executives accountable for their systemic criminal activity. With oversight responsibility and the ability to legislate tougher laws for the pharmaceutical industry, one would hope that Congress would get involved and fix these issues. This may be harder than it would seem, however, based on the significant influence and power the industry wields through their lobbying efforts and political donations. In a study of the pharmaceutical industry, Dr. Olivier J. Wouters of the London School of Economics and Political Science, found that between 1999 and 2018, Pfizer spent $219 million in lobbying expenses and $23 million in campaign contributions.[26] With avarice and potential corruption hindering a congressional fix to the problem, we are left fighting for our rights with little to no assistance.

Throughout 2020 and continuing to this day, media entities appear to want average Americans to fear the COVID-19 virus. By contrast, the lack of media attention paid to the pharmaceutical industry's criminal activity, lobbying efforts, and political donations, appears to be an indication that media entities are desperate to get American citizens to trust these untrustworthy companies. Government officials, for their part, chose to prioritize speed over both good scientific methods and the safety of their own people. The subsequent campaign of fear played a significant role in the decision many made to receive a COVID-19 vaccine. Many Americans responded by fearing the virus either for themselves or on behalf of a vulnerable loved one.

As COVID-19 vaccine mandates began to roll out in 2021, many Americans who would not have chosen to get the vaccine based on their own risk assessments did so due to fear of losing their jobs. The fear, apparently missing in many Americans' assessment of whether to take the COVID-19 vaccine, was a healthy and reasonable fear of the possibility of adverse reactions from products produced by untrustworthy pharmaceutical companies. These companies are motivated by maximizing profits, which usually has an inverse relationship with maximizing safety. Also missing for many Americans was a healthy fear of their own government, with its long history of trampling rights and a possible connection to the creation of the pandemic through NIH funding.

All this fear, targeted at individual health, is not a modern phenomenon. Dr. Benjamin Rush, a member of the Sons of Liberty and a signer of the Declaration of Independence, understood well the role that fear played for medical patients in his own era. In a 1789 lecture to medical students on the duties of a physician, Dr. Rush described how unethical physicians disgraced his profession by stoking the fear of danger or death to increase the likelihood that patients would accept or look favorably upon the prescriptions being offered by those physicians. When their patients died under their care, these unethical predictions provided an escape from blame. When the full truth is known in our current situation there will be no escape from blame for those who unethically stoked the pandemic of fear. Ultimately, over 265 million Americans have chosen to take at least one dose of a COVID-19 vaccine. The pandemic of fear played a significant role in convincing so many Americans to participate.

CHAPTER 5

The Military's Response to a Near-Peer Virus

I hope also that the recent insults of the English . . . will establish the eternal truth that acquiescence under insult is not the way to escape war.[1]

—Thomas Jefferson

The Department of Defense's initial reaction to the outbreak of COVID-19 in Wuhan, China, was relatively quick. In a January 31, 2020 Fact Sheet, the DoD noted that the Commander of US Indo-Pacific Command had restricted all DoD travel to the People's Republic of China (PRC) in support of a US Department of State Level 4 Travel Advisory. The DoD also participated in an evacuation of State Department employees, dependents, and US citizens from Wuhan, China, and provided continued monitoring of those nearly 200 individuals for follow-on care and medical transport if required.[2] By March of 2020 all official DoD travel, both domestically and abroad, had been restricted with limited exceptions for things deemed mission critical. Even service members who wished to take leave were required to remain in the local area.[3] Commands throughout the military were directed to establish programs for COVID-19 testing, contact tracing, patient isolation, and quarantine measures for DoD personnel potentially exposed to COVID-19.[4]

In a March 27, 2020, message to all DoD Personnel, then Secretary of Defense Mark Esper laid out the DoD's priorities in the face of the

COVID-19 pandemic. He listed protecting DoD personnel and their families along with supporting President Trump's "whole-of-nation response"
to the pandemic. He explained the reasons behind the various mitigation
measures and provided guidance to commanders to ensure a healthy and
mission-ready force. He also empowered subordinate commanders to make
their own health mitigation decisions to ensure mission accomplishment.
"I trust our commanders around the world," he stated, "to make the best
decisions for their troops as they balance mission requirements with force
health protection."[5] The most important thing I think Secretary Esper did
early in the pandemic was to keep the DoD focused on national security
while balancing the need for a healthy force. He understood our adversaries
and knew they would be looking for opportunities to take advantage of any
US defense weaknesses. He stated, "Our adversaries may look to exploit this
crisis, as much of the world's attention is directed toward the coronavirus.
We will not hesitate to modify our security posture around the world, if
necessary."[6]

The announcement of Operation Warp Speed by the Trump administration turned the nation's attention toward the hope that the pandemic
could be eradicated by the rapid development of a vaccine. President Trump
gave a Rose Garden press conference about Operation Warp Speed on May
15, 2020, with Secretary of Defense Mark Esper and the chairman of the
Joint Chiefs of Staff, General Mark Milley, in attendance. While President
Trump's remarks were an occasion of great hope for many, some concerning
things were presented at that press conference, including the fact that the
administration would "cut through every piece of red tape" for the sake of
speed. President Trump announced that scientists had begun working on
the first vaccine candidate on January 11, 2020, and that he wanted to have
an effective vaccine by the end of the year. He also noted that, typically,
pharmaceutical companies have to wait to manufacture a vaccine until it
has received all of the necessary regulatory approvals and that this can delay
public availability. President Trump went on to announce that "our task
is so urgent that, under Operation Warp Speed, the federal government
will invest in manufacturing all of the top vaccine candidates before they're
approved." President Trump admitted there would be risks stating, "It's
risky, it's expensive, but we'll be saving massive amounts of time."[7]

Many Trump administration officials remain unapologetic to this
day about permitting Operation Warp Speed to throw out the normal
safety processes and vaccine development timelines. In his 2022 memoir, Mark Esper spoke glowingly of Operation Warp Speed, calling it an

"overwhelming success."[8] While many would consider it a reckless safety risk, Esper boasted about Operation Warp Speed, stating that "We reduced a process that normally takes five to ten years down to around eight months. And we not only met our time goal, we also doubled our success with *two* different vaccines [Pfizer and Moderna]."[9] Secretary Esper also did an odd thing in his memoir, calling Operation Warp Speed a "second Manhattan Project endeavor —one that would deploy the full powers and resources of the federal government to bring vaccines into production faster than any of the experts thought possible."[10] While Esper may have compared Operation Warp Speed to the Manhattan Project as a way to highlight the speed of the government program, the similarities, ultimately, may not end there.

The Manhattan Project, a secret government program to rapidly develop nuclear weapons, has a complicated legacy. The dawn of the atomic age did usher in a new energy source, nuclear power, capable of sustained energy production unimagined before 1945. However, the Manhattan Project was first successful as a weapons program, which resulted in the deaths of hundreds of thousands of innocent Japanese citizens. Is the reckless speed at which our government developed a new and novel vaccine technology also going to cost a significant number of lives before it can be harnessed?

At the time of the May 15, 2020, press conference, it was well known that the virus had originated in China. The Wuhan Institute of Virology lab leak theory was already being discussed publicly. President Trump had previously fielded questions about the lab leak theory from reporters during a press briefing the month before.[11] With the focus on Operation Warp Speed and the race to create a COVID-19 vaccine, President Trump was again asked about China. As one of the reporters inquired, "What happens if China is the country that develops the vaccine. . . . Will the US still have access to that vaccine?" President Trump's answer was essentially, "yes."[12] Despite President Trump's answer, the question was an insightful one in light of the possibility that China, one of our most significant strategic adversaries and the country in which SARS-CoV-2 originated, would look to exploit the pandemic to gain strategic advantages over us.

Following the announcement of Operation Warp Speed, the Department of Defense appeared to increasingly focus on the COVID-19 virus and the various mitigation measures meant to combat it. COVID-19 was quickly becoming the primary "enemy" our nation's combat force was fighting. Even the language being used by some of the most senior military leaders was shifting as they began attributing various human characteristics to the virus. In a June 23, 2020, message to all forces under the command of 4-star admiral Christopher Grady, Commander of US Fleet Forces, the COVID-19 virus was called a "continued threat to mission assurance" that could "infiltrate our forces." The message went on to state that "Asymptomatic spread is a reality and one misstep opens a potential attack vector for this virus."[13]

Ultimately, the vaccine was the tool that the DoD planned to utilize to eradicate this "continued threat to mission assurance." When the first COVID-19 vaccine received an Emergency Use Authorization on December 11, 2020, the military was ready. They had already built a COVID-19 Vaccine Distribution Plan and Population Schema.[14] Implementation was nearly immediate. Following the lead of various political leaders, commanders up and down the chain of command began encouraging service members under their charge to get vaccinated. In a February 26, 2021, message to all Navy medicine personnel, Navy Surgeon General Rear Admiral Gillingham, called the vaccine "our biological body armor" and "our safest and best weapon against this unrelenting adversary." [15] Continuing to proselytize for the vaccine he stated, "By simply rolling up your sleeve you can don this amazing biological body armor, make a personal contribution to community immunity and project medical power."

The language placing the COVID-19 virus at the pinnacle of US military threats was not just rhetoric. Commanders were required to report every outbreak and every positive case. Every close contact was tracked. Command vaccination rates were tracked by higher headquarters, and commanders whose units lagged behind had to answer tough questions. By early 2021 our entire military force had turned from an outward focus on national threats to an internal focus on individualized health. It is likely that this shifting of focus placed us at significant strategic risk from the national threats we were previously paying much closer attention to.

On March 4, 2021, the new secretary of defense, Lloyd Austin, issued a memorandum for all Department of Defense employees in which he laid out his priorities, similar to the memo from his predecessor, Secretary Esper, a year earlier. Secretary Austin organized his priorities into three main categories: Defend the Nation, Take Care of Our People, and Succeed Through Teamwork. Listed second within the "Defend the Nation" priority was "Prioritize China as the Pacing Threat." Listed first, before the "Prioritize China" effort, was "Defeat COVID-19."[16] It was ironic that "Defeat COVID-19" was not listed under the "Take Care of Our People" priority. That it was listed under "Defend the Nation" and listed *before* the "Prioritize China" effort, is indicative of the mis-prioritization our national security strategy seemed to take in 2020 and 2021. Many of the newest efforts, policies, and programs seemed to be based on political motives rather than national security. We appeared to lose focus on our strategic rivals in an effort to promote certain new ideologies and to refashion the COVID-19 virus into some sort of new "unrelenting adversary" that intends "to infiltrate our forces."[17]

On July 29, 2021, President Biden directed the DoD to investigate how and when to "add COVID-19 vaccination to the list of required vaccinations for the military."[18] Secretary Austin responded with an August 9, 2021, memo to all DoD employees in which he laid out his intentions regarding the COVID-19 vaccine as well as a proposed timeline. Secretary Austin cited public reporting suggesting that the Pfizer-BioNTech vaccine would achieve full FDA licensure in the next month. Referencing this timeline, he stated that he would mandate the vaccine immediately upon full licensure or seek a presidential waiver "no later than mid-September."

The August 9, 2021, memo was also significant because it laid out Secretary Austin's apparent intention to begin discriminating against unvaccinated service members and federal workers. After laying out his timeline for mandating the vaccines, he announced that he would "comply with the President's direction regarding additional restrictions and requirements for unvaccinated Federal personnel."[19] Just in case any service members missed the subtlety of what appeared to be tacitly approved coercive measures, he added, "Those requirements apply to those of you in uniform as well as our civilian and contractor personnel."[20] At this point in time, mask mandates, lockdowns, travel restrictions, contact tracing, and involuntary quarantines had been in place for service members for nearly a year. Alluding to "additional restrictions and requirements for [the] unvaccinated" immediately after announcing an upcoming vaccination mandate could only be interpreted as coercive measures against the unvaccinated. To me the language from the August 9, 2021, memo felt like I was being told *get vaccinated, or we will make your life unbearable.* Commanders who later got aggressive in coercing and discriminating against unvaccinated service members had a reasonable argument in pointing to Secretary Austin's August 9, 2021, memo as the authorization for their actions.

Not every senior leader was on board with the impending vaccination mandate for the military. There were significant strategic risks in implementing a DoD-wide mandate to use a new mRNA technology that had only been available to the public for less than a year. The claims that these products were "safe and effective" had not been proven. The human trials of these products were still underway and would not be completed until 2023.

There were also serious questions about long-term effects since there was no way to replicate time in any trial. The only way to test for long-term effects would be to let enough time pass for any long-term side effects to become manifest.

One senior officer who publicly aired his concerns was Navy Commander Jay Furman. In an August 15, 2021, article, Furman laid out scientific research about the vaccines and the risks to the largely young and healthy military population. He warned that mandating the COVID-19 vaccine could "trigger manning shortfalls brought on by resignations and lost enlistments from this all-volunteer force."[21] He also noted that "Pressing forward against these extremely large unknowns by mandating COVID-19 vaccines could potentially threaten basic military deployment assumptions, to say nothing of the long-term destruction to morale and recruiting."[22] Commander Furman's most important recommendation was to ensure that half the force remain unvaccinated as a risk-mitigation strategy. Due to the significant unknowns regarding safety, efficacy, and long-term side effects, maintaining half the force as unvaccinated would guarantee that at least one of the two groups would remain a viable fighting force regardless of any long or short-term vaccine side effects.[23] The long-term effects of vaccination could prove to be debilitating, or even deadly, at statistically significant rates. A properly and strategically focused Operational Risk Management analysis should have led DoD leaders to demand a control group for the COVID-19 vaccine in order to ensure a healthy segment of the force.

The necessary DoD risk analysis was either not completed properly or the results were ignored for the sake of political expediency. When the FDA fully licensed the first COVID-19 vaccine under the trade name Comirnaty® on August 23, 2021, the DoD was again ready. The next day, Secretary Austin issued a memorandum mandating the COVID-19 vaccine to all Department of Defense service members. As will be discussed at length in later chapters, the order to be vaccinated included the statement that "Mandatory vaccination against COVID-19 will *only* use COVID-19 vaccines that receive full licensure from the Food and Drug Administration (FDA), in accordance with FDA-approved labeling and guidance."[24] This order and the language Secretary Austin used aligning his order to federal law would have significant ramifications, particularly in light of the subsequent actions many military leaders later took that violated Secretary Austin's order.

Within the Navy, most service members who were not already vaccinated against COVID-19 were given a printed-out version of the order to sign so that the Navy had documentation that the service member had received and acknowledged the order. When I was handed my order to be vaccinated, the same order handed to many thousands of service members, it included seven statements, some of which were very controversial, and several of which I was convinced were false official statements. I was tracking the overwhelming attention the Navy was paying to the COVID-19 virus. I was deeply concerned about how we were anthropomorphizing the virus and seemingly placing the virus itself above all other threats on our list of adversaries. Historically, national security and defense of our nation had always been the Navy's number one concern. So, I assumed that the very first statement on the order I received, "Your health and safety are the Navy's number one concern,"[25] was a false official statement, because the alternative meant that a number of commanders at the highest levels were committing a dereliction of duty that could actually threaten our national security.

In a response I provided my chain of command, I informed them that this statement was patently false and an outlandish contradiction of Navy Doctrine. If the individualized preservation of personal health implied by this statement were true, we would never put sailors in harm's way. The over 1.1 million American service members, who shed blood and died defending our nation, stand in stark contrast to our present leadership's assertions that an individual's health is the number one concern of our military. Those fallen heroes would be shocked to see how we have acquiesced to a disease and replaced their highest concern, defending the Constitution unto death, with many leaders' professed new number one concern, preserving each service member's individual health and safety.

We know for a fact that our adversaries, especially our near-peer adversaries, try to take advantage of any weaknesses to further their own national interests. The fact that the COVID-19 virus originated with one of those adversaries should have played a more significant role in how we viewed that adversary. We should have responded with much greater caution. Instead, we played into the hands of that adversary and almost immediately replaced that adversary with the virus itself on our list of "threats." By turning inward, we lost focus on the most significant threat and provided our near-peer adversaries the perfect opportunity to exploit our weaknesses. Thomas Jefferson warned us that acquiescing in the face of threats would not make us safer. A war with a near-peer adversary will not be avoided by acquiescing in the face of things that make us weaker. Our absurd mis-prioritization of individual health preservation above actual nation-state threats, rather than making us more ready for war, has likely hastened the day our adversaries feel confident enough to make that war a reality.

CHAPTER 6

Accommodating Equal Rights of Conscience

The civil rights of none shall be abridged on account of religious belief or worship . . . nor shall the full and equal rights of conscience be in any manner, or on any pretext infringed.[1]

—James Madison

The right to freely worship was of tantamount importance to our Founding Fathers. Looking at the totality of their writings, the argument can be made that freedom of religion was the first principle upon which our nation was founded. Individual rights related to religious freedom were placed as prominent cornerstones of both the Declaration of Independence and the Bill of Rights. James Madison, one of the architects of the First Amendment, had fought hard to ensure that "toleration" was not the cited basis for individual religious freedoms. He argued that the exercise of faith was an inalienable right and not a gift from the government as the word "toleration" implied. His argument won the day, and the final language of the First Amendment is one of the great legacies we have of James Madison's efforts on behalf of religious freedom.[2]

Fast forward nearly 240 years and we find ourselves in a situation where our government has set itself at odds with the free exercise of religion and has started to operate as if religious freedom is something granted by the state. The COVID-19 pandemic has seen significant expansions of government power into the free exercise and free expression rights of individuals.

Service members are often the most vulnerable when they become the targets of intended government programs. Service members, however, do not give up basic constitutional rights when they take their oaths of office to join the military. Service members still have the right to free speech and the right to the free exercise of religion. The law is clear and applies to all Americans including service members (with limited restrictions for activities while in uniform). In accordance with the Religious Freedom Restoration Act, Title 42 U.S.C. § 2000bb, in general, a government *shall not* substantially burden a person's exercise of religion. The only exception permitted by law requires the government to demonstrate to the burdened individual that the government has a compelling governmental interest for overriding their rights and that the actions taken are the least restrictive required to further that compelling governmental interest. The Department of Defense Instruction 1300.17 implements this law for the military and provides guidance on how the services are to accommodate the religious practices of their service members. DODINST 1300.17 "Religious Liberty in the Military Services," provides procedures for accommodating the religious practices of service members, and prohibits adverse actions and discrimination based on sincerely held beliefs.

On September 15, 2021, just three weeks after the Secretary of Defense's mandate, I submitted a religious accommodation request to be exempted from the requirement to be vaccinated against COVID-19. The basis of my religious accommodation request was threefold. First, each vaccine available had some connection to aborted fetal cells, either through testing or through their use in the vaccine itself. Second, the COVID-19 vaccine mandates were a violation of the moral principle that non-morally obligatory acts must be voluntary. Lastly, these mandates violated the moral principle of therapeutic proportionality.

Abortion is such a politicized issue in today's society that I became convinced that my chain of command would be dismissive of my religious beliefs on the matter. What they likely did not know was that the abortions performed to produce fetal cells for vaccines are done on human babies whose hearts are still beating when the genetic material is extracted.[3] Having a beating heart is so important to producing usable cells that a device known as a Langendorff apparatus was developed to keep a heart beating even after it has been extracted. Since it is nearly impossible to find a detailed explanation of how to prepare human subjects for the Langendorff apparatus, a 2017 review of the procedure for mice is useful to understand how it works.

In the article, the author describes how it is important to keep the animal free from stressful stimuli and recommends protocols of anesthesia. Once calm, the animal is place in the supine position and the diaphragm is cut using a trans-abdominal incision to expose the thoracic cavity. The beating heart is removed and placed in an ice-cold chemical solution to rinse off the blood. This will temporarily stop the heartbeat and is a time period that must be minimized. Finally, the heart is connected to the Langendorff apparatus by the aorta and a nutrient rich oxygenated solution is pumped through the heart. The heart will start beating again within seconds after being connected to the Langendorff apparatus.[4]

The above description was provided as part of a study of mouse hearts. However, we have evidence that the same Langendorff techniques were used to study and extract material from human hearts as well. A 2012 Stanford study of human cardiac progenitor cells revealed that the source of the human hearts being studied was a company called StemExpress and that a Langendorff apparatus was used to procure the "human fetal hearts" in question.[5] A 1999 article in the *Pediatric Research* journal provided much greater detail about the human subjects being studied. The article describes a study of six human hearts that were extracted from human beings at between 18–22 weeks of gestation.[6] The author's note that the "fetal hearts were aseptically obtained after elective termination of normal pregnancy by dilation and evacuation . . . [and t]he hearts, weighing 3–5 g each, were immediately dissected from the thoracic cavity with the great vessels intact and transported to our laboratory within 15 min."[7]

It is unconscionable to me that living human beings, who happen to still be in the womb, are not protected as persons under the Fourteenth Amendment. In light of how much more we now understand about the science and gestation of human life, I believe this element of the law is ripe for review by the courts. Regardless of fetal personhood, if these babies are evacuated from the womb with their hearts still beating, they are considered born and are protected by Section 1 of the Fourteenth Amendment which states, "All persons born . . . in the United States . . . are citizens of the United States and of the state wherein they reside. No state shall make or enforce any law which shall abridge the privileges or immunities of citizens of the United States; *nor shall any state deprive any person of life*, liberty, or property, without due process of law." If abortion practitioners are following the standard procedure for use of the Langendorff device by vivisecting living human beings outside the womb, they are committing murder under every law we currently have and under the current understanding of the Fourteenth Amendment. I did not learn all of this until much later in life, but once I did, I knew that I could never participate in abortion any way, including by taking vaccines tainted by abortion. The vivisection of a non-consenting living human person is something that can never be tolerated, nor justified, regardless of when the procedure happens or what benefits may come from it. The moral principles rejecting such practices were well established in the Nuremberg Code.[8]

Even if my chain of command disagreed with my strongly held religious convictions with regard to abortion, I was convinced that every member of the military and every citizen of our nation should be very concerned about the therapeutic proportionality of the COVID-19 vaccines. My own beliefs require me to apply the principles of therapeutic proportionality to all medical decisions. Therapeutic proportionality, simply stated, is the assessment of the benefits of a medical intervention in light of the risks. For example, if someone gets a scratch on their arm, the moral principle of therapeutic proportionality precludes preemptive amputation of the arm to avoid the very small risk of developing gangrene, which left untreated, could potentially cause death. Applying this moral principle to vaccination requires us to study everything about these vaccines, the short- and long-term effects of these vaccines, and who stands to benefit from mass vaccination (financially or otherwise). Contributing to my own therapeutic proportionality analysis of the COVID-19 vaccines were the alarming safety signals that were already present in the Vaccine Adverse Event Reporting System (VAERS), a vaccine injury data collection program cosponsored by the CDC and FDA. I included this information in my religious accommodation request. I also included publicly available data demonstrating that the actual efficacy of the various COVID-19 vaccines was far below the 95 percent advertised by the pharmaceutical companies that produced them.

All of this was present and understood by those of us researching it as early as September 2021. Since that time, the data has continued to confirm the poor safety profile and the poor efficacy of every single COVID-19 vaccine. Overall, for a young and healthy person, as I was, the COVID-19 vaccines completely failed to meet basic therapeutic proportionality requirements. If the risks of a medical treatment outweigh the benefits, or if there are no benefits at all, such as giving COVID-19 vaccines to the young and healthy, forcing someone to take that medical treatment is a violation of their basic human rights in addition to being a violation of therapeutic proportionality.

Military members across all services also had beliefs and convictions that precluded them from receiving a COVID-19 vaccination. Because the First Amendment protects all religious beliefs and even protects the dictates of conscience for those with no religious beliefs, the number and type of religious accommodation requests received by the military departments were significant. Receipt of a COVID-19 vaccine made no sense for many, particularly the young and healthy. The overwhelming propaganda pushing for mass vaccination trumpeted highly questionable information and sometimes outright falsehoods as facts. The underlying politicized environment was actively trampling individual rights under the pretext of public health. I continue to be amazed that there were not more religious accommodations requests filed. Many service members had concerns, but out of fear, misplaced trust, or a misdirected sense of duty (to superiors as opposed to the Constitution) they went along with the mandate.

There are many of us still studying the various issues while trying to understand and make sense of these complicated factors. Even to this day we continue to find questionable circumstances surrounding both the pandemic as well as the vaccines developed to "end" it. This was the environment in which I, and thousands of my fellow service members, filed religious accommodation requests to be exempted from taking the COVID-19 vaccine. Little did we know at the time, but these requests would fall on deaf ears.

For the Navy, authority to approve COVID-19 vaccination accommodation requests was delegated to the Chief of Naval Personnel, Vice Admiral John Nowell. Each O6 (Navy captain), or higher ranking commander, had the responsibility to forward any religious accommodation request they received to the Chief of Naval Personnel with a recommendation to approve or disapprove. I trusted my leadership to follow the law and their own instructions by giving my request a thorough and individualized review. Because I understood this section of the law and because I had thoroughly read the implementing regulations, I knew that the Navy had to either accommodate my religious beliefs or prove a compelling government interest that outweighed my beliefs. My friend Matt Tennis, a submarine officer and Catholic father to eight children, had recently submitted a religious accommodation request of his own. His commander, a Navy rear admiral, had recommended approval with a simple justification: "Forwarded, recommending approval based on no operational impact to the command." This recommendation for approval had given me great hope that the Navy was intending to honor the religious convictions of service members and approve requests such as Matt's and my own.

My hopes received a devastating blow when my own commanding officer wrote an elaborate three-page memorandum to Vice Admiral Nowell recommending disapproval of my own religious accommodation request. I found out later that he had received help from our higher headquarters Judge Advocate General (JAG) lawyer, who coached my commanding officer on the ideal key words and phrases that would give the color of law to a disapproval of my request. What confused me the most about the disapproval recommendation from my commanding officer was the fact that he completely ignored the extensive evidence I provided that supported my therapeutic proportionality convictions. He recommended disapproval, but did so without even attempting to prove the compelling government interest that would outweigh my religious convictions. Therefore, I could not understand the reasoning behind recommending disapproval. The only thing I could come up with was that he might be using me, and my religious accommodation request, as an opportunity to show the chain of command how aligned he was with the vaccination program.

If there was anything telling about this episode of the pandemic for me, it was my own naiveté with regard to the trust I still placed in my chain of command and in Navy leadership at the highest levels. The Navy had already played their hand a year earlier and had demonstrated their disregard for individual religious freedoms. In a June 23, 2020, lockdown order from US Fleet Forces under the command of Admiral Christopher Grady, service members were directed to only travel to work, home, and for specific essential business: "food, medical, pharmacy, gas, and child care services." The order took the extra step of listing a significant number of off-limits locations and activities. These locations and activities included swimming pools, gyms, barber shops, cinemas, dine-in restaurants, concerts, public beaches, and indoor religious services. The addition of indoor religious services was shocking and immediately struck me as a violation of the Constitution.

I was then directed to sign this order so that my chain of command could verify that I had read and understood it. However, the order restricted my constitutional right to worship. As previously noted, to be a lawful order or to even have the inference of lawfulness, the order must not be contrary to the constitutional rights or the statutory rights of the recipient. I knew that an order denying my right to worship at Mass was not one I could follow in good conscience. Yet the top of the document I was asked to sign stated, "I will abide by the following." I knew it would also be wrong of me to lie and sign the document if I did not intend to stay away from indoor religious services. I therefore asked for the editable version of the document, removed the words, "I will abide by the following," and replaced them with, "I acknowledge the following." I was essentially telling my chain of command that I acknowledged receiving the order, but that I could not abide by it.

As a senior officer, I understand well the potential consequences of violating an order, especially if the one giving that order believes it to be lawful. Section 5.A.7.B.4.H of Admiral Grady's order stated that "This is a lawful general order under Article 92 of the Uniform Code of Military Justice (UCMJ). Any person subject to the UCMJ who violates or fails to obey this order may be subject to adverse administrative and/or disciplinary action." While I disagree that this was a lawful order, I was and continue to be willing to accept all the consequences of following my faith, just as I am willing to accept all the consequences of defending the Constitution against those who violate it.

I was not the only one who found this order to be a violation of the Constitution. According to a well-placed O6 source, "there was an O6 uprising after the Fleet Forces order." A significant number of O6s provided ample feedback up their chain of command that the order prohibiting indoor religious services had gone too far. It was also reported to me that two-star Admiral John Meier, Commander of Naval Air Forces Atlantic, had issues with the order. In a meeting he called with his subordinates shortly after the order was issued, Admiral Meier stated that he considered the order only a tool for commanding officers should they feel it necessary in the course of their command. "To be clear," he said, "I draw your attention to the words 'may be used' and submit that I have no intention of punishing anyone for violating these orders."

I admire this level of conviction from anyone, but to have it from someone so senior is unique in the current leadership environment. Most of our current crop of strategic and policy level leaders seem unwilling to risk their careers to do the right thing or to even take care of the subordinates they lead. That was not the case for Admiral Meier in this situation. Even if Admiral Meier did not personally consider the order to be unlawful or unconstitutional, he was still willing to stand in the gap for his subordinates who did feel that way and whose beliefs precluded them from following it.

Due to the significant outcry, the Trump administration got involved and Admiral Grady was forced to permit attendance at indoor religious services just over two weeks after the original ban.[9] Acting Undersecretary of the Navy Gregory Slavonic directed the Navy and Marine Corps service chiefs "to inform Commanders to incorporate this clarification allowing attendance at religious services."[10] Regardless of the positive outcome for service members of faith, this action was a tip of the hand for how the military would treat religious convictions when the COVID-19 vaccination mandate was issued a year later. I also believe Admiral Grady's order opposing the free exercise of religion was a contributing factor in his continued career progression.

CHAPTER 7

Constitutional Tap Code

As our number grows less, let us love one another proportionably more.[1]

—Benjamin Franklin

When service members are caught behind enemy lines and taken prisoner, one of the methods used to break their spirits is to isolate them. Communication can be a great comfort and one of the most important mechanisms for keeping morale high in the face of significant abuse. Denying communication and human contact can make a prisoner more malleable by removing those things that can anchor them to their beliefs and to each other. The American prisoners of war held captive by the North Vietnamese, had developed a unique solution to maintain communication when normal communication methods resulted in severe punishment when caught. These heroes developed a "tap code" by which they communicated to each other in ways their captors could not decipher. This communication was their lifeline and morale booster. It was also what gave many of them the strength to endure through some of the toughest challenges they would ever face.[2]

Following the military vaccine mandate, commanders began instituting measures to coerce service members into receiving COVID-19 vaccines. The fact that many of these service members had religious convictions that precluded them from receipt of a COVID-19 vaccine was not a hindrance to many commanders. Service members who persevered in their convictions and remained unvaccinated were often the victims of efforts by their commands to isolate them and keep them from experiencing camaraderie within their military units.

Many commanders measured their success by the number of subordinates they could get to cave to the COVID-19 vaccination order. The tactics used to enforce the vaccination order ranged from simply informing service members of the order all the way to coercive and discriminatory abuse to compel compliance. It did not matter to some commanders that many of the unvaccinated service members in their charge had submitted religious accommodation requests. If they were the type of commander to apply coercive pressure, they were also likely to ignore any religious beliefs that conflicted with enforcing the order.

An analogy is useful to demonstrate why this is wrong. Let's say an order was given that every service member had to eat pork, and one of your subordinates announced he was Jewish and it was against his faith. Would it ever be OK to apply coercive pressure to make that subordinate break with his belief system and eat pork? What about an order to drink alcohol directed towards a Muslim? It would be morally wrong in both cases. Leadership should support subordinates with religious beliefs and find some way to accommodate those beliefs. It is not just the right thing to do morally and ethically, it is also the law, and every commander has an obligation to follow the law.

In the case of religious objections to the COVID-19 vaccine, commanders up and down the chain of command did not respect the professed beliefs of subordinates who informed them they could not receive the vaccine. The law states that the military should accommodate those beliefs or prove a compelling reason why they are unable to do so. Most commanders did not even bother paying lip service to the rights of their subordinates and did not wait for this process to play out. Instead, they began devising ways to get service members to deny their beliefs and take the COVID-19 vaccine.

The reports of discriminatory and coercive tactics were many. Some commanders granted early Friday afternoon liberty for all who had received the COVID-19 vaccine but required all those who had not received it, including those who had filed religious accommodation requests, to attend Friday afternoon musters (a personnel accounting process typically involving a roll call and inspection) at medical. In the Navy, some commanding officers granted liberty out in town to vaccinated sailors, but restricted unvaccinated sailors to the ship during port calls both in the United States and overseas. Many pulled unvaccinated service members from the work they were trained for and gave them menial tasks below their qualifications and training levels. These actions were clearly punitive as they often had no impact or relationship to unit health or readiness. In many cases, these actions decreased readiness rather than improved it.

Not every service member who wanted to put in a religious accommodation request was permitted to do so. The story of Seaman Recruit Owen demonstrates the Navy's subtly aggressive denial of constitutional rights within the Recruit Training Command at Great Lakes. Seaman Recruit Owen signed up to join the Navy well before the vaccine was mandated for the military. He came in under the Delayed Entry Program which allowed him to finish school, work through the summer, and then report for duty in late fall of 2021. When Secretary of Defense Lloyd Austin mandated the COVID-19 vaccine for the military on August 24, 2021, Owen reached out to his recruiters to inform them he could not take the vaccine for religious reasons. His recruiters told him that he could request a religious accommodation upon arrival at boot camp. They assured Owen that his country still needed him, but they did not help him file a religious accommodation request, nor did they provide him with information on how to start the process.

When Seaman Recruit Owen showed up at Recruit Training Command, in Great Lakes, IL, he went through the same two-week quarantine that every unvaccinated recruit was forced to endure. The unvaccinated recruits were split into "pods" of eight individuals, separated by squares marked with tape on the ground in the large bunk rooms. They were then ordered to remain inside their tape-marked squares. The quarantine was much more like a prisoner-of-war confinement camp than the normal boot camp. The recruits were fed and provided study materials; however, Owen reported receiving only a single one-hour outside recreation period during the entire two-week quarantine. After the quarantine was over, they were sent to medical for basic screening and for vaccines. Seaman Recruit Owen, along with several other recruits, declined the COVID-19 vaccine. He also asked to file a religious accommodation request citing his recruiter's assurance that he would be permitted to do so. His request was ignored.

He and the other unvaccinated recruits were given "counseling" by leadership and the medical staff, who berated them for refusing the vaccine and lectured them about its safety and efficacy. One of the medical "counselors" told Owen that "people like you are the reason we're still in this pandemic." His continued refusal earned him additional isolation. While the vaccinated were moved on to the next phase of training, the remaining unvaccinated recruits were isolated and kept from follow-on training phases. Leadership often pulled the unvaccinated aside for additional lecturing and abuse, threatening them with disciplinary actions, including separation from the military. The fact that many recruits had religious beliefs that precluded them from receiving the COVID-19 vaccination was completely ignored. DoD Instruction 1300.17, which precludes adverse action, discrimination, or denial of training based on the expression of sincerely held beliefs, was also completely ignored. Seaman Recruit Owen asked to file a religious accommodation request multiple times and was denied. He asked to speak to a Navy Chaplain, not knowing that a chaplain interview was the first step in filing a religious accommodation request. In violation of numerous Navy regulations and basic constitutional rights, Owen's request to speak to a Navy Chaplain was completely ignored.

Instead of helping Seaman Recruit Owen file a religious accommodation request, speak to a Navy Chaplain, or grant him any support, he was isolated, discriminated against, berated, and threatened. The Navy's unlawful actions eventually caused nearly every unvaccinated recruit in that class to cave and receive the COVID-19 vaccination. Owen, however, remained true to his beliefs and his conscience and continued to hold out in the hope he would eventually be granted the right to serve. Instead, his lieutenant summoned him for additional interrogation and abuse. Owen again requested help from his lieutenant to file a religious accommodation request. The lieutenant denied his request, continued to threaten him, and ultimately began the process of removing Seaman Recruit Owen from the Naval Service. Owen served honorably for thirty-nine days before he was finally sent home, his constitutional rights denied and his religious beliefs unsupported.

I was able to debrief Owen nearly nine months after his unceremonious removal from the Navy. After hearing his story, I thanked him for his thirty-nine days of honorable service. I also let him know that according to regulation all service members who have served honorably after September 11, 2001, are awarded the National Defense Service Medal. No one at Recruit Training Command told Owen about the National Defense Service Medal he had rightly earned, nor did they likely appreciate the truly honorable way that Owen had defended the Constitution in the face of unconscionable coercion. Following our conversation, and in recognition of his courage and integrity, I presented Owen with a National Defense Service Medal that I had personally worn. Seamen Recruit Owen did more to defend the Constitution in his thirty-nine days of honorable service than any of the military leaders who put their consciences and oaths aside to advance their careers by promulgating harmful politically expedient orders.

In the first months after the military vaccine mandate, my personal feelings of isolation were significant. Other than a few close friends who shared my own values, I could not find any other service members who had similar concerns about the COVID-19 vaccines. We were operating in a low-information environment where the "safe and effective" rhetoric was pushed with incredible aggression. Yet, the data supporting the "safe and effective" assertion was not available. We had to seek out that data independently, analyze it ourselves, and share it as best we could. The first military voice I found sharing real data and experiences that countered the overwhelming narrative came from Lieutenant Colonel Theresa Long.

In September 2021, Lieutenant Colonel Theresa Long was serving as the Brigade Surgeon for the 1st Aviation Brigade, stationed at Ft. Rucker, Alabama. In addition to extensive experience as an Army Medical Officer, she had received specialized training in emerging infectious disease threats, medical management of chemical and biological casualties, and aviation safety. In a September 24, 2021, affidavit, Dr. Long exposed the results of her own study of the COVID-19 vaccination risks and benefits for the military population in light of the requirement for a mission-ready force. Dr. Long noted that the military-aged population under study had a 99.997 percent survival rate after contracting the SARs-CoV-2 virus. She also noted that there were significant risks related to the administration of the COVID-19 vaccines. She offered a number of conclusions, most importantly that none of the COVID-19 vaccines "can or will provide better immunity than an infection-recovered person."[3]

Finding Dr. Theresa Long was a significant moment for me at this point in my journey. I had felt a bit like a lone wanderer, lost in the dark and unable to see the treacherous cliffs on either side of me. Encountering Dr. Long's public opposition to the prevailing narrative was like seeing a shining light ahead and hearing the voice of a friend guiding me to safety. I will forever be grateful for her courage in standing up when so many did nothing and said nothing about the questionable things happening around them. I found Dr. Sam Sigoloff and Dr. Peter Chambers a short time later. Dr. Sigoloff is an Army major and board-certified family physician. Army Lieutenant Colonel Pete Chambers, now retired, served for thirty-nine years as a highly decorated Green Beret and Army flight surgeon. Like Dr. Theresa Long, Dr. Sigoloff and Dr. Chambers were outspoken in their attempts to inform service members of the true science behind the vaccines. Both were subsequently removed from their positions.

Before his retirement, Pete Chambers had been the surgeon responsible for the medical readiness of four thousand soldiers.[4] In accordance with both his Hippocratic Oath and his oath to the Constitution, Dr. Chambers executed his legal obligation to provide informed consent briefs to the soldiers in his charge. His brief, packed with scientific information and supporting data, was long but incredibly effective at adequately informing his soldiers of all possible risks and benefits of the COVID-19 vaccines.[5] As a result of the extensive scientific information passed to them by Chambers, only six of the four thousand soldiers elected to be vaccinated. The politicized and career-focused leaders above Lieutenant Colonel Chambers were furious. The informed consent brief was ruining his leaders' vaccination percentages "and therefore their chances at promotion."[6] When Major General Charles Aris, Commanding General of the 36th Infantry Division, confronted Chambers, Pete attempted to explain the science behind the vaccine informed consent briefings. General Aris then slipped up and revealed a truth many of us would not discover until much later. In response to Chambers' scientific explanation, General Aris responded, "It's not about the science but about the politics."[7]

Dr. Long, Dr. Chambers, and Dr. Sigoloff were instrumental in early efforts to discover and reveal scientific concerns about the COVID-19 vaccines to service members. These three whistleblower doctors were also some of the first military medical experts to make the connection between the vaccines and the likelihood of significant warfighting readiness risks. In being courageous champions of medical science, they also exposed the politicized nature of so many senior leaders who refused to conduct a legal and constitutional analysis of their orders as they are required to do by their oaths. The impact these military doctors had on me was instrumental in convincing me that we had to begin building a network of support for others struggling through their own isolation, discrimination, and abuse.

Little did I know but networking efforts were already underway. My friend Mark Zito, a Navy lieutenant commander and fellow Catholic I had known for more than a decade, introduced me to many of the early military vaccine information and support networks. Networking came naturally to Mark, and he had already begun building a group of like-minded military contacts. He also consistently went above and beyond the call of duty to personally help me whenever I asked for anything. He was the reviewer, preparer, and redactor of much of my early written work in opposition to the military vaccine mandate.

Mark also connected me to Air Force Major David Beckerman, who was the driving force behind a growing network focused on providing a resource exchange platform for service members. It was within this group that I got to meet and work with some of the most brilliant minds and incredible leaders the military has to offer. Some of those included Air Force Lieutenant Colonel JJ McAfee, Navy Commander Olivia "Liv" Degenkolb, Navy Commander John Sharpe, Marine Corps Major Patrick Henry, and several military JAG lawyers. This group eventually expanded to hundreds of service members. In the early days, however, the group was much smaller, and we divided the research responsibilities, shared new information with each other, reviewed each other's written works, and worked to expand our network of constitutional defenders.

JJ McAfee, one of the more senior members of our group, was recently selected for promotion to full-bird colonel in the Air Force. With the systemic discrimination against unvaccinated service members occurring across the military, we were concerned that the Air Force would deny his promotion. The Navy had just done a similar thing to their unvaccinated population. In an October 13, 2021, message to the fleet from Admiral William Lescher and Vice Admiral John Nowell, commanders were directed to "delay the promotion of any officer refusing the vaccine," and "withhold the advancement of any enlisted member refusing the vaccine."[8]

In a move that struck me as ironic, the message withholding advancements and ordering promotion delays called the vaccine mandate a "lawful order" three different times. Both the CNO and SECNAV messages from late August, ordering service members to be vaccinated with a COVID-19 vaccine, also used the words "lawful order" twice each.[9] Nearly every official message on the subject, across all branches of the Department of Defense, attempted to hammer home that the order requiring vaccination against COVID-19 was a "lawful order."

Throughout my naval career, I don't ever remember orders being so insistent that they were "lawful." I have received hundreds of orders since 2003, including Operations Orders, Fragmentary Orders, Warning Orders, Temporary Duty Travel Orders, and Permanent Change of Station Orders. Orders were just orders. They rarely needed to confirm to recipients that they were actually lawful orders. Even my most recent Permanent Change of Station orders, in which I was directed to uproot my family and transfer to a new command hundreds of miles away, did not once use the phrase "lawful order." The overuse of the phrase "lawful order" seemed designed to overcome the justifiable hesitancy and religious convictions of service members. In fact, its use was so prevalent and insistent that it seemed even senior leaders at the highest levels were subconsciously trying to convince themselves this order was somehow lawful.

One of the best outcomes from the growing network of like-minded service members coming together to support each other through the various hardships was the organic launch of numerous local chapters of support. As unvaccinated military members found each other, those in close proximity began to meet to provide much needed support and camaraderie. Most were excluded from participating in their own command's functions and social routines. The various local chapters of unvaccinated support stepped in to fill the gap. We held "hails and farewells" for unvaccinated service members who had been banned from their own commands. A "hail and farewell" is an unofficial military function in which new members of a unit and their families are welcomed to the team and departing members are given a proper send-off with all the requisite speeches and inside jokes about their time with the unit. We welcomed new service members and their families as we found them. We also held events for those who were allowed to retire and those who were forced to separate involuntarily.

One of the first retirement send-offs I had the honor of participating in was one we held for Navy Musician First Class Drew Stapp. Drew had served for twenty-two years and had retirement orders in hand to be executed in December of 2021. He went ahead and moved his family to Texas in August 2021 so that his children could start school. Then the mandate dropped, and Drew submitted a religious accommodation request citing religious convictions that precluded his receipt of a COVID-19 vaccine. Drew was subsequently removed from his position, denied any meaningful work by his command, and essentially banned from coming to work. When December rolled around, Drew was denied the right to execute his retirement orders. There was no logical reason, or any health risk to the Navy, that could justify keeping him from traveling to his family and completing the already approved retirement orders.

Those of us who were local got together and supported Drew and others in similar unjustifiable situations over the next several months. It was not until the courts forced the Navy to do the right thing that Drew, after eight months of separation from his family, was finally permitted to complete his retirement and travel to Texas to be with his family. Our group of local unvaccinated service members gave Drew a proper retirement send-off thanking him for his twenty-two years of honorable service.

Another local group at Camp Lejeune, North Carolina was able to provide support to one of the Marine Corps' early administrative separation victims. An administrative separation in the military is the non-judicial process by which the military can involuntarily discharge a service member without giving them the opportunity to demand a trial by court-martial where they could invoke their Fifth and Sixth Amendment rights. The Marine Corps was more aggressive than many of the other services in attempting to separate unvaccinated service members as fast as possible.

A federal court could stop the Marine Corps by ordering a halt to adverse actions through a preliminary injunction. A federal court in Florida eventually ordered a preliminary injunction halting adverse Marine Corps actions against the unvaccinated, but this order did not come soon enough to save the career of Kim Warren (a pseudonym). Kim, a Marine Corps Lance Corporal, had her religious accommodation request denied and her command moved rapidly through the subsequent administrative separation processes. Upon completion of the administrative separation process, her command ordered her to pack her things. They picked her up at her barracks and then unceremoniously dropped her off outside the gate of the base. They did not even bother taking her all the way to the airport. The local unvaccinated support network organized a group to meet her at the gate and provided her a proper send-off deserving of her strength and courage in the face of tyrannical coercion. They gave a few short speeches to thank her for her service, prayed together as a group, and provided her a ride to the airport.

The spirit of isolation was prevalent nationwide throughout the lockdowns of 2020. The military continued the isolations in 2021 and 2022 but targeted unvaccinated service members almost exclusively once the COVID-19 vaccines were mandated. Unvaccinated service members were not subjected to torture like many American prisoners of war. However, human communication and human contact are both so critical to mental well-being that efforts to isolate and coerce unvaccinated service members essentially amount to cruel and unusual punishment. Like the heroic American prisoners of war isolated from each other by the North Vietnamese, unvaccinated service members had to come up with unique and creative solutions to contact and support each other. Defending the Constitution requires us to maintain human contact and communicate our love for one another using whatever method, or "tap code," we can.

CHAPTER 8

The Nuremberg Shrug

It is remarkable, that the officers and soldiers of our enemies, are so totally depraved, so completely destitute of the sentiments of philanthropy in their own hearts, that they cannot believe that such delicate feelings can exist in any other. . . . But in this they are mistaken, and will discover their mistake too late.[1]

—John Adams

From the earliest days of the military vaccine mandate many of us had a strong suspicion that there was more wrong and unlawful about the mandate than immediately met the eye. The problem we had was few of us were legal experts; almost none of us had ever before read the various laws and military regulations that govern the administration of FDA-licensed products and products approved by the FDA for emergency use. These two types of biological products, I quickly learned, were two very different things. I was very blessed at this point in my journey to be surrounded by a growing network of military leaders who took to researching the law as if their lives and livelihoods depended on it, which, ironically, they did. We were also connected to several brilliant legal minds with decades of experience dealing with the military justice system, including John Sharpe, Tom Rempfer, R. J. Lopez, Dale Saran, Major Candace White, Lieutenant Commander Kristi Bao, Lieutenant Commander Pat Wier, and Captain Carmel Tomlinson. The combined skills and experience of this team was able to put us years ahead of where we would otherwise have been if we were left to study and fight this legal battle on our own.

Most of what we learned about the law in this area does not just apply to the military, but to every single American citizen. The only law unique to the military is 10 USC §1107a, which permits the president to waive certain requirements for military personnel before a product can be administered. Other than this one exception for military personnel, everything about legal rights that I will lay out in this chapter apply to all Americans, and the government cannot remove, waive, or withhold these rights. If you are an American citizen and you are reading this, these laws apply to you and can be used to protect you against future mandates of Emergency Use Authorized products.

Following the August 24, 2021, DoD vaccine mandate from Secretary of Defense Lloyd Austin, the service secretaries and service chiefs moved quickly to promulgate COVID-19 vaccine orders to their subordinate commands. The speed of the subsequent Navy vaccine orders was a telling indication of the aggressiveness this mandate would receive from certain segments of leadership. Secretary of the Navy Carlos Del Toro issued his order to the Naval Service (Navy and Marine Corps) to be vaccinated against COVID-19 on August 30, 2021.[2] His immediate subordinate, the chief of Navy operations, Admiral Michael Gilday, issued his vaccination order to the Navy a day later on August 31, 2021.[3] In something of a leadership and communications faux pas, Admiral Grady, commander of US Fleet Forces, beat his supervisors to the punch by issuing his own vaccine mandate order on August 28, 2021, days before his superiors. Admiral Grady was apparently so eager to issue his own vaccine mandate for the significant number of forces under his command that he did not wait for the normal trickle-down progression of orders. Again, when it came to politically expedient issues, Admiral Grady found a way to stand out from his peers.

It is important to note that each of these orders only mandated vaccines that had been fully licensed by the FDA. Each of these orders also permitted service members to fulfill their COVID-19 vaccination requirement with a product that had received an Emergency Use Authorization (EUA) from the FDA, *but in each case the directive explicitly stated that the receipt of an EUA product was voluntary.* Since I had already resolved not to receive the COVID-19 vaccine because it would violate my conscience, I was curious about the legality of the order and decided to go to my local medical treatment facility to see if the FDA-approved product, named Comirnaty, was actually available. On September 7, 2021, I confirmed in person with my medical provider that they did *not* have the FDA-approved product, Comirnaty. Hundreds of service members across the nation took similar actions to confirm that Comirnaty could not be found at any military medical facility or at any civilian pharmacy or medical provider.

Why was this action on our part so important? At the most fundamental level, Americans never lose the right to legally refuse an EUA product. The law controlling the use of EUA products, 21 USC § 360bbb, Authorization for Medical Products for Use in Emergencies, imposes significant responsibilities upon the government to inform Americans of their rights, including the right to refuse. The only exception to the government's duty to inform citizens of their rights is in the narrowly defined presidential waiver process for the military per 10 USC §1107a. This exception only waives the required condition that service members be *informed* of their right to refuse an EUA product. Congress passed 10 USC § 1107 into law, in part, because of the myriad injuries sustained by service members and veterans due to past military medical abuses including atomic testing, Agent Orange, Persian Gulf War Drugs, and vaccines.[4] This was quickly followed by the passage of 10 USC § 1107a, which specifically addressed use of EUA products. Similar to the constitutional violation of failing to provide a suspect with their Miranda rights, not informing a potential recipient of their right to accept or decline an EUA product, either by presidential waiver or by omission, does not remove the underlying rights protected by statute and the Constitution.

Prior to receiving an EUA product, the recipient must be informed of the option to accept or refuse the EUA product, as codified in 21 USC § 360bbb-3(e)(1)(A)(II)(iii). This right is a required condition that the Secretary of Health and Human Services (HHS) "shall" include for the authorization of any unapproved product covered by an emergency declaration. This means that by law no one can mandate EUA products, and the government must inform recipients of their right to refuse. This law covers all types of EUA products including test kits, masks, and COVID-19 vaccines. It may be surprising to many Americans to know that not a single mask or COVID-19 test kit has received full licensure from the FDA. The FDA has only authorized masks[5] and test kits[6] through Emergency Use Authorizations. This means that every single American, regardless of the type of mandate, has had (and continues to have) a legal right to refuse masks and COVID-19 tests.

The military services quickly realized they had a problem with their inventory of FDA-approved COVID-19 vaccines. Basically, there was no inventory of the approved product, and therefore no legal way to mandate receipt of the millions of doses of EUA product still on the shelves. Shortly after the order mandating the vaccine, Pfizer posted an announcement on the NIH website that they would not produce any of the licensed product "over the next few months while EUA authorized product is still available and being made available for US distribution."[7] Military leaders did not slow down the implementation of the vaccine mandate to wait for a product that could be legally mandated. I'm sure none of them wanted to go "hat-in-hand" to the administration, letting their executive branch political leaders know they had jumped the gun. They elected to place political expediency and their own careers first and did not stand up for the legal and constitutional rights of the service members in their charge. Instead, they acquiesced to what appeared to be a purely profit-driven decision by Pfizer, while attempting to come up with a clever work-around to make an unlawful action appear lawful.

The Navy surgeon general, Rear Admiral Gillingham, authored what appears to be the very first memorandum claiming interchangeability of the EUA product with the fully licensed product. Rear Admiral Gillingham's memo, dated September 3, 2021, states that "The FDA-approved vaccine, and the vaccine used under the EUA, have the same formulation, and can be used interchangeably to provide the COVID-19 vaccination series without presenting any safety or effectiveness concerns. Navy medical providers can use Pfizer-BioNTech doses previously distributed under the EUA to administer mandatory vaccinations."[8] Mr. Robert Hogue, Acting Assistant Secretary of the Navy for Manpower and Reserve Affairs, signed a similar memorandum only five days later on September 8, 2021, claiming that "Navy medical providers can use Pfizer-BioNTech doses previously distributed under the EUA to administer mandatory vaccinations."[9] Finally, Assistant Secretary of Defense for Health Affairs (ASD HA) Dr. Terry Adirim, wrote a September 14, 2021 memorandum stating "these two vaccines are interchangeable and DoD health care providers should use doses distributed under the EUA to administer the vaccination series as if the doses were the licensed vaccine."[10]

These three memoranda take medical advice from the FDA and use it as the basis for stripping service members of their legal right to decline receipt of an EUA product. Dr. Adirim specifically cites the FDA's Q&A website to justify use of EUA Pfizer-BioNTech vaccines in lieu of the FDA-approved Comirnaty.[11] The FDA's Q&A website provides medical advice regarding the use of the EUA product to complete a "vaccination series" but does not, and cannot, make a legal determination about the use of *forced* or *mandated* EUA products. The FDA website did not address the *legal* difference between the products, nor was it a determination of biosimilarity or

interchangeability, which has specific statutory requirements per 42 USC §
262(k)—Licensure of Biological Products as Biosimilar or Interchangeable.
The law cites critical requirements for interchangeable products, including
that:

1. a sponsor must submit an application for licensure of the
    biosimilar product;
2. both products become fully licensed before being declared
    interchangeable, and;
3. "Approval of an application under this subsection [Licensure of
    Biological Products as Biosimilar or Interchangeable] may not be
    made effective by the Secretary until the date that is 12 years after
    the date on which the reference product was first licensed under
    subsection (a)."

In accordance with federal law 42 USC § 262(k), no product may be legally
declared interchangeable with Comirnaty until at least *August 24, 2033*! As
further evidence, the FDA's authoritative source for approved biologics, the
"Purple Book," lists "no interchangeable data at that time" for Comirnaty[12]
How these three DoD officials thought they could circumvent the law and
get away with providing military leadership a plausible excuse for unlawful
actions continues to baffle me.

Every military commander who cited interchangeability as the justification for their unlawful orders ignored the legal distinction between the
two products. Most notable of these legal distinctions is that one was a
non-available licensed product and the other an available EUA product.
The available EUA product imposes a requirement on the administrator
to inform recipients of their inherent right to refuse. This legal distinction
was clearly cited by the FDA in every Pfizer BioNTech and Moderna EUA
reissuance letter since full licensure.[13] The law governing the administration
of EUA products was completely ignored by the military.

Thousands of service members across the DoD, including me, brought
this information up to leadership. We demonstrated that the fully licensed
product was unavailable and that the available product was an EUA product. Leadership was informed that interchangeability did not apply and
could not convert an unlicensed product into a licensed product. All of this
information was ignored, prompting more aggressive methods of getting
leadership's attention. Service members began filing internal complaints,
congressional inquiries, and Inspector General complaints. Leadership

dismissed these filings and refused to engage the arguments or even refute the facts we laid out. Leaders at the highest levels decided that the policy would continue and there would be no discussion, regardless of the policy's questionable merits or the overwhelming evidence that its implementation was violating multiple laws.

In addition to ignoring the information we provided, military leadership began threatening adverse action, administrative separation, and even courts-martial. What was so disturbing was that these threats were being made against people whose only offense was following their consciences and exercising their right to refuse EUA products. We were not criminals violating the law or the rights of others. We had not violated any historical standard of performance or military requirement. The military requirement that we were accused of violating was a COVID-19 vaccination mandate that had been created mere months earlier. It was unproven, untested, and was so obviously an important political agenda for the current administration that we were astounded that senior military leaders were not willing to sacrifice their own stars and careers to protect the integrity of the services they led.

We were left incredibly confused and disheartened that there was no support to be found from almost any leader, including from those who continued to repeat the mantra that "we have to put people first." Where was the accompaniment or basic human pity for even the junior service members who had no one else to turn to? Instead, many of our leaders continued to push the politically expedient agenda that would advance their own careers rather than protect the rights of those they led. I felt the words J. R. R. Tolkien wrote following his experiences in The Great War [World War I] were particularly apropos to this situation: "to him that is pitiless the deeds of pity are ever strange and beyond reckoning."[14] Instead of compassion, pity, or even delicate feelings, we faced the heartless mantra that "I'm just following orders." My own commanding officer, not understanding my faith or my decision to refuse the COVID-19 vaccine, told me of his hope that I would get the vaccine and be able to continue serving. I responded that "I am willing to give my life for my country. What sort of man would I be if I wasn't even willing to sacrifice my job for my faith?" I am sure my response was equally incomprehensible to him, but it was the only stance I—and so many others—could take in good conscience. Regardless of our petitions, responses, or complaints, it appeared that commanders at all levels were abdicating their responsibility to stand up for truth, the law, or the Constitution.

In the Navy this leadership phenomenon within the junior officer ranks is often referred to as the "ensign shrug." The ensign is the lowest-ranking regular officer in the Navy. They typically have no experience and are often just months out of college. Since they are officers, many expect them to know something, to be able to give some guidance, or to simply direct the actions of those put in their charge. This is rarely the case at first. A seasoned deck-plate leader, typically a chief petty officer, will usually take them under their wing, and the chief and the commanding officer of their unit will guide their training. Until the ensign has gone through this development and gained some experience, they often have no idea what they are doing or how to do it. When confronted with the myriad things they do not yet know, they give the proverbial "shrug" and ask for help. Sometimes this is caused by fear of the massive bureaucracy they have attached themselves to and sometimes the "shrug" is due to being overwhelmed with everything that they still have to learn.

A variation of this phenomenon occurred at all levels of the chain of command during the rollout of the DoD COVID-19 vaccination campaign. In this case the "shrug" was not due to a lack of knowledge. Leadership was informed by thousands of service members about the unlawfulness of their actions. The "shrug" was also not due to a lack of experience because these commanders had spent entire careers gaining the experience required to reach whatever level of leadership they were at. The "shrug" in this case was due to an inexplicable combination of trust and fear.

Officers up and down the chain of command, including some of the most talented and capable ones, resorted to blindly trusting leadership above them rather than investigating the concerns we brought to their attention. This was also a time of great fear in the military. Fear of deviating from the guidance of superiors played a part, as did fear of losing station, stalling careers, and being removed from positions of command and authority. We were left with leaders at all ranks relentlessly repeating *"I am just following orders,"* the same excuse Nazi war criminals used to justify their actions at the Nuremberg trials in 1946. The tone it was delivered in was most often a resigned one, but it was delivered nonetheless. This *"just following orders"* response from our commanders was so incessant during the military COVID-19 vaccination campaign that I have come to call it the "Nuremberg Shrug."

CHAPTER 9

The Best Defense
Is a Good Offense

*If we had nothing to fear from any offensive operations of the enemy,
policy may require very extensive and important offensive operations
on our part.*[1]

—General George Washington

By October 2021, what had originally been a trickle of religious accommodation requests became a flood. Thousands of service members spent tens of
thousands of hours researching and writing their requests to be exempted
from the COVID-19 vaccination requirement. We did this monumental
amount of work under the assumption that our leadership would read and
review our requests. The growing support network played a key role in sharing research and helping those whose commands were either providing zero
support or were actively undermining the service members in their charge.
We sincerely believed that our depth of research would help convince our
chains of command not only of the sincerity of our religious beliefs, but also
of the very serious risks to force health and force readiness that we clearly
foresaw.

Small things continued happening that caused our outlook to darken
regarding the support we would get from leadership for our religious beliefs.
In a discussion we had before I submitted my religious accommodation
request, my commanding officer stated that he did not think any religious
accommodation request would ultimately be approved, and he could not

justify having unvaccinated sailors on missions. My commanding officer also told me I had a "moral obligation to be vaccinated to protect the most vulnerable." When my commanding officer ultimately recommended disapproval of my request, I was devastated but, unfortunately, not very surprised. What was more alarming, however, was some of the language he used in his disapproval recommendation.

I did not know until much later that my commanding officer built his ostentatious three-page disapproval recommendation with the assistance and guidance of our higher headquarters JAG lawyer. The disapproval recommendation attempted to cover every possible justification permitted by DoD Instruction 1300.17 for denying a religious accommodation request. The underlying assumptions made by my commanding officer for many of his justifications were largely indefensible. For example, my commanding officer stated that military readiness would be seriously degraded if I remained unvaccinated. This statement flew in the face of the fact that I had helped lead our command through the entire pandemic with *zero* degradations to military readiness.

My performance was a significant factor in maintaining our military readiness. In fact, during my tenure as executive officer, the conscientious approach we took to the health and safety of our sailors since the beginning of the pandemic caused our command to have exactly *zero* transmissions of the COVID-19 virus at any of our facilities, events, or trainings. I did all of this while remaining unvaccinated. It was nonsensical to imply that refusing the COVID-19 vaccine was suddenly seriously degrading our command's mission readiness. The assertion that, by being unvaccinated, I was degrading mission readiness was disproved by the evidence of the previous eighteen months.

Due to my own apprehensions about the future, largely brought on by the intense opposition of my own commanding officer, this period of my journey became a time of increased personal fear. A significant number of service members in our network had been reporting that they were being threatened with courts-martial in addition to other adverse administrative actions. Those who were already being separated were being given separation paperwork citing a "commission of a serious offense." Charging a service member with the commission of a serious offense has historically been reserved for the most heinous of crimes, but was now being applied to refusers of the politically-charged order to be vaccinated. There were very serious concerns that the DoD would try to make examples out the more senior of us and those of us who had been more public about our opposition to the mandate. The DoD could attempt to court-martial us for the "commission of a serious offense." A court-martial conviction for a serious offense comes with the very real possibility of jail time. It was already obvious to us that the DoD was violating the law, both by mandating EUA products and by violating First Amendment religious freedom rights. There was little doubt in my mind that, if they thought they could get away with it, the DoD would probably try for just such a conviction to make an example of us to others.

This led to some very serious conversations with my wife. How far was I willing to go for what I believed in? St. Thomas More became my model and inspiration as I chose my next steps. His unwillingness to violate his conscience and go along with the intended actions of King Henry VIII led to his persecution and eventual martyrdom. With my wife's encouragement, we decided that I would follow St. Thomas More's example and follow my conscience wherever it led. We also had great family support that provided something of a cushion if the worst happened. Together, my family and I put a contingency plan in place to cover the possibility that I would be convicted and jailed for following my conscience. My wife would take our seven children to live with family while I served my time. Upon release, I would find whatever work I could to provide for the family, even if it meant swinging hammers or digging ditches.

This discernment process was one of the most freeing experiences I have ever had. I gave my fear to God, and He returned to me a joy and peace about the future I had not anticipated. Those of us who believe in heaven and hell and attempt to live our lives such that we end up in the former and not the latter are often given the Grace to no longer cling so desperately to this life. This Grace allows us to have a much more eternal focus as we make decisions regarding our temporal lives. Thousands of service members were making similar decisions and placing their honor, their faith, and their eternal souls above their earthly concerns. Several months after my own experience of giving this situation to God, I had the great honor of speaking with Lieutenant Colonel C. Scott "Sonny" Duncan, USMC. Sonny is a fighter pilot, a former TOPGUN instructor, and commanding officer of an F-35B fighter squadron. Sonny shared with me that he and his family went through a nearly identical journey and came to the same decision, that he would serve jail time rather than break with his faith. Finding Sonny and so many like him gave me great courage. I was consoled by the knowledge that whatever I went through, I would not go through it alone.

As our network of service members grew into the hundreds, we begin tracking the growing number of religious accommodation request disapprovals. During this time, we did not see a single approval. For the Navy, the ultimate adjudication authority, Vice Admiral Nowell, was using his common access card (CAC) card to sign the disapproval documents. When a CAC card is used to sign a document it places the signer's information, including their DOD identification number, into the signature block along with a timestamp. As the disapprovals began pouring in, we knew something was wrong. There were way too many disapprovals happening each day for someone to actually be giving our requests an honest review. Many of our requests were heavily researched and were quite lengthy. The award for the longest request I personally know about came from Lieutenant Commander Mark Zito who submitted a request of nearly four hundred pages. In an effort to figure out exactly how much time Admiral Nowell was spending with each of our requests, we began gathering the disapprovals and tracking the timestamp of Admiral Nowell's signature.

Our network had a few hundred members, but over four thousand Navy service members had filed religious accommodation requests. Since we did not have all the data, it would be difficult to figure out what was going on with our religious accommodation requests within Admiral Nowell's staff. It was a massive understatement to say that transparency was not a priority for them. Despite our best efforts, I struggled to find ways to prove that Admiral Nowell and his staff were violating the law and not doing the individualized compelling interest analysis required for each request. We could clearly see what was happening, but proving that their (dis)approval process was breaking the law to a level that would hold up in court was a bigger challenge. Our big breakthrough occurred through the networking efforts of Lieutenant Commander Matt Tennis.

Matt had recently come in contact with an individual working on Admiral Nowell's staff. A discussion ensued and the staffer expressed some deep concerns with what he was seeing at the Navy's Accommodation Review Team, the team responsible for adjudicating all of the Navy's religious accommodation requests. Matt pulled this individual into our network and made introductions. Due to his unique position in the "belly of the beast" and his resolve to uphold his oath to the Constitution, I almost immediately began referring to him as "Captain Courageous" to those close to the situation. Captain Courageous began sharing actions with our group that he believed clearly violated both military regulations and the Constitution. Captain Courageous was a unique outlier and a hero to us.

One of our number even told him, "You're a great American. I hope my grandchildren one day know your name." His response was, "I'm nothing special. I'm a normal person trying to do the right thing. As far as I'm concerned, this is part of my oath. My rights were given to me by God; they are not for sale. They weren't given to me by the government, and the government doesn't have the right to take them away."

Thanks to the front-row seat provided by Captain Courageous, we knew that the Accommodation Review Team was struggling to keep up with the massive influx of religious accommodation requests. On November 3, 2021, one of the Accommodation Review Team leaders called for "an emergency meeting." In an email to the rest of the staff, this leader said, "I have a major announcement that requires your presence, if you are unable to attend please call me this afternoon for back briefing." At that meeting, the rest of the Accommodation Review Team was told they would now be processing one hundred religious accommodation requests a day. According to Captain Courageous, the Accommodation Review Team was also told that this effort was being done "so that they know it will stand up in court."

From a different whistleblower we had already received reports that JAG lawyers at the highest levels had informed senior Navy leaders that the Navy was "taking on significant litigation risk" by systematically denying all religious accommodation requests. We asked Captain Courageous how accelerating the review process and increasing the number of requests denied each day would enable the process to "stand up in court" better. His response was that the acceleration would *not* help the process stand up better in court but that was the line they felt the team needed to hear. "It would be a pretty big blow to morale," he said, "to tell people that they are breaking the law in the same conversation where you are telling them they have to work nights and weekends." Apparently the hundred religious accommodation requests a day processing quota was not enough for leadership. Just two days later, on November 5, 2021, the Accommodation Review Team was asked to process 150 requests a day.

Based on the information provided to us, it was apparent that senior JAG lawyers were already thinking about and preparing for litigation over this very politized issue. The most senior JAG officer in the Navy, and the one most responsible for ensuring that Navy policy did not violate the Constitution, was Vice Admiral Darse Crandall. As the Judge Advocate General of the Navy, Admiral Crandall was the principal military legal counsel to the secretary of the Navy and the chief of naval operations, Admiral Mike Gilday. Even if Admiral Crandall did not consider it important to keep the military from becoming politicized, he did have an obligation to ensure that commanders were following the law and not violating the Constitution.

Instead of prioritizing the defense of service members' constitutional rights, it became quickly obvious that, under Admiral Crandall's leadership, the Navy JAG Corps was circling the wagons in defense of policy regardless of the legality of the orders being issued. Essentially, the JAG Corps was enabling potential unlawfulness and violations of constitutional rights. The inside information we received from Captain Courageous confirmed our suspicions that enforcement of policy was prioritized by the JAG Corps above their own sworn obligations under the Constitution and as officers of the court.

The most significant item that Captain Courageous publicly released was the disapproval standard operating procedure (SOP), written by Admiral Nowell's staff to streamline the systematic denial of all religious accommodation requests. Hundreds of religious accommodation request denials personally signed by Admiral Nowell were pouring out of his office, and we now knew exactly what process was used to produce those denials. The disapproval SOP was a written instruction detailing exactly what an Accommodation Review Team member was to do when receiving a religious accommodation request. The procedure had six phases with a total of fifty steps. Each step was exhaustively detailed, describing exactly what the team member must type and each button they must click as the requests are received, denial letters prepared, and all documents routed for review. There were even computer screenshots of nearly every step of the process so there would be no confusion.

The very first processing step required of Admiral Nowell's staff was to place the requester's name in the "to:" line of the disapproval template and to rename the disapproval template file to include the last name, first name, and rank of the religious accommodation requester. It was hard to believe when we first read it, but preparing the disapproval template was done before the actual request was even required to be opened. The disapproval letter was then placed with the request and only then sent for review by various Navy administrative offices, including the Special Assistant for Legal Matters (the "official" JAG review of each request). Once the reviews were completed by the required offices, an internal memo was prepared for Admiral Nowell requesting him to "sign . . . letters, *disapproving* immunization waiver requests based on sincerely held religious beliefs."

Only after all the above steps were completed, and the disapproval of the religious accommodation request was a foregone conclusion, did the disapproval SOP finally direct the team member to actually open it, read through it, and list any pertinent details. There was a note in all caps on this particular step which emphasized to the team member that this review was important to maintaining the fiction that religious accommodation requests were receiving a case-by-case examination: *"This is the most critical step in the entire process and the CNO [Admiral Gilday] and CNP [Admiral Nowell] are relying on you to ensure that your review is thorough and accurate. Do not rush this process and ensure that you understand before moving forward."*

This entire process, and especially the "critical" step towards the end, was a blatant violation of the Navy Personnel Instruction 1730.11a. This instruction, titled *Standards and Procedures Governing the Accommodation of Religious Practices* states that each request must be reviewed "on a case-by-case basis, giving consideration to the full range of facts and circumstances relevant to the specific request," and "requests to accommodate religious practices should not be approved or denied simply because similar requests were approved or denied."

I found it ironic that this step was considered so critical to Admiral Gilday and Admiral Nowell. It wasn't critical for understanding our religious convictions or in weighing those convictions against a compelling government interest, because this was not done before drafting the disapproval letter. The disapproval "fix" was already in. No, this step was critical in protecting Admiral Gilday and Admiral Nowell from lawsuits for violating the Religious Freedom Restoration Act and for violating our constitutional rights.

It appeared to me that using the final steps of the process to create a spreadsheet of details from our religious accommodation requests was meant to provide "proof" that a review was done in case they were ever audited or sued. We still do not know what level of involvement Vice Admiral Crandall personally had in developing, reviewing, or approving the process that was trampling the rights of Navy sailors. Regardless of his personal involvement, as the Judge Advocate General of the Navy he had a legal obligation to ensure the constitutional rights of service members were protected and that the Navy was following the law in reviewing religious accommodation requests. This legal obligation was not fulfilled and the rest of the JAG Corps followed suit. Despite the significant evidence of a systematic violation of First Amendment constitutional rights, the JAG Corps was apparently prioritizing the legal defense for senior leaders like Admiral

Nowell and Admiral Gilday, above their own obligations to ensure that Navy policies were lawful.

The disapproval SOP was the proof we were looking for. It demonstrated that those who were processing our religious accommodation requests, and Admiral Nowell who denied all those requests, were committing numerous violations of military regulations and constitutional rights. The courage displayed by Captain Courageous in getting the disapproval SOP out to the public would play a significant role in later lawsuits and in stemming the tide of unlawfulness. Captain Courageous was not the first military whistleblower during the pandemic and his release of a document was seemingly a small thing. However, as is often the case throughout history, it is the small things that make the most important victories possible.

Reading the disapproval SOP and seeing the proof that the law and my own constitutional rights were being violated significantly shifted my mindset toward those violating the law. Individuals, organizations, or even entire governments that violate constitutional rights with impunity must be opposed. Our leaders were ignoring our valid concerns and all the evidence we provided that unlawful actions were occurring. I realized, as did a number of our network, that we had to go on the offensive if we were going to have a chance of stopping the lawlessness. So many of our leaders wrongfully trusted that the process was lawful and that subsequent disciplinary actions were justified. We were surrounded by leaders who either actively attempted to dismantle the Constitution or silently let it happen in front of them, unopposed. I certainly felt like I was behind enemy lines. I began to think of the entire situation as an operational battle-problem that had to be solved. I had given the outcome to God and was no longer particularly concerned about personal consequences, provided that I continued following the law and my own conscience. This decision gave me significant freedom to maneuver as I debated how to begin lawful offensive operations against enemies of the Constitution.

CHAPTER 10

Betrayal, Exposure, and Illness

I shall pay the strictest attention to my orders, but cannot flatter myself with effecting as much with raw militia . . . and I am very apprehensive that the troops, by being exposed, will become sickly, and desertions follow in consequence.[1]

—Major General Benedict Arnold to
General George Washington (One month
before being exposed as a traitor.)

Both military regulations and the law provide a number of mechanisms that can be used, even by the most junior of service members, to identify unlawfulness being committed by superiors and to get it corrected. Service members can go directly to their superiors with issues and concerns, or they can file a complaint to the Inspector General of their service. If service members feel they have been wrongfully discriminated against they can file an Equal Opportunity complaint. If internal efforts gain no traction, service members can go to Congress and request a congressional inquiry. US law 10 USC § 1034 protects service members and their communications with members of Congress. The law does not permit military leadership to interfere with a service member who desires to speak to their member of Congress or retaliate against them for this communication.

Ultimately, service members still have the right to free speech and can use that right to bring public light to issues that need to be remedied. In many ways, this book is part of an ongoing effort to exercise my own First Amendment rights and to help make the public aware of a very serious wrong occurring in the military over an unlawfully implemented vaccine mandate. While it is certainly not the only current military issue that should be concerning to the American people, it is the one I was most intimately involved with.

As for lawful mechanisms for fighting back against wrongs, my personal favorites are the UCMJ Article 138 complaint process and the Navy Regulations Article 1150 complaint process. UCMJ Article 138 allows a service member who feels they have been personally wronged by the actions of their commanding officer to seek a remedy, also known as a redress, of those wrongs. UCMJ Art 138 requires that the service member first request the desired redress from the commanding officer. If the commanding officer denies that redress request, they can file their complaint of wrong to the general court-martial convening authority over the commanding officer. The general court-martial convening authority is typically the first general or admiral in that commanding officer's chain of command.

The Navy Regulations Article 1150 process is unique to the Navy and very similar to the UCMJ Article 138 process. The main difference between the two processes, and one that would play a critical role in later court cases, was that the Article 1150 process could be filed against any superior whom a service member felt had wronged them. The Navy Article 1150 process also did not require a prior redress request before filing the formal complaint, making it an easier process to initiate, and did not give the recipient a heads-up that a formal complaint could be coming. One of the other great benefits of either the Article 138 or the Article 1150 in the Navy is that if they are filed against a three- or four-star admiral, the complaint is automatically forwarded to the Naval Inspector General as required by SECNAV Instruction 5800.12C, *Investigation of Allegations made against Senior Officials of the Department of the Navy*. This benefit allowed any complaint against a three- or four-star admiral to double as an inspector general complaint as well. From my perspective, this doubled the efficiency of such complaints.

By late November 2021, I had filed more than half a dozen requests for redress to my commanding officer and to various O6s in my chain of command highlighting concerns about the vaccine mandate and potential violations of First Amendment rights. These official requests for redress did not include the many informal discussions and emails I had over these issues with leadership. Most of my questions and concerns were ignored. Of the responses I did receive, not a single one actually engaged on the merits of my concerns nor offered any evidence or insights that would make the military's vaccine guidance defensible.

The only unique response I received was from one of the O6s in my chain of command. I had asked for redress for a false official statement he made about the COVID-19 vaccines being "safe and effective." I provided significant evidence in my request that contradicted his "safe and effective" statement. Instead of countering with evidence or discussing the merits of the COVID-19 vaccine, he wrote me a memo reiterating that "As the Secretary of Defense stated [in his memo of 24 August 21], the mandatory vaccination is safe, effective and necessary to defend America." He also recommended that I "review [UCMJ] articles 88, 89, and 133 (contempt toward officials, disrespect toward superior commissioned officer, and conduct unbecoming an officer and a gentleman, respectively)."

I do not know how this O6 intended me to receive his guidance to review these three punitive articles of the UCMJ, but I certainly saw a veiled threat in it. In a humorous twist, though, the O6 went too far and tipped his hand by including UMCJ Article 88 in his response. Upon further review (as I was directed to conduct), I found that in order to actually prosecute someone for Article 88, the accused must use contemptuous language against the "President, the Vice President, Congress, the Secretary of Defense, the Secretary of a military department, the Secretary of Homeland Security, or the Governor or legislature of any State." I had, of course, used absolutely no contemptuous language against anyone in my request. Nor did my request mention a single one of the required institutions, officers, or individuals. I had already committed to following both the law and my conscience regardless of the consequences, so I was not intimidated. Either way, his response made it clear that relief would not be found in the ranks of field-grade officers or with senior command leadership.

Although many Americans may not realize it, the military actually receives its highest level of strategic guidance and operational direction from civilian leaders, not from senior uniformed service members in the general and flag officer ranks. The president of the United States is the commander-in-chief of the US Armed Forces and appoints civilian secretaries to lead the DoD and each of the armed services. Directly under the civilian leadership is the policy and strategic level of uniformed military leadership consisting of mostly three- and four-star generals and admirals. Although it should not have happened, political leaders began to use the military to further political aims several decades ago, and this trend has only accelerated in recent years.[2] These political leaders' subordinates at the three- and four-star levels are left in a very tough position, determining when to push back on guidance from civilian political leaders and when they should go so far as to resign in protest to protect the services they lead from the corrosive consequences of a politicized military.

Vice Admiral Dean Lee, USCG (Ret), described this particular problem very eloquently in a recent radio interview. He described the training that the most senior military officers go through after they put on their third star. The program, likely one of the most exclusive training programs in the world, is called Pinnacle. Pinnacle brings in some of the most brilliant and powerful people to talk about the most important strategic national security concerns and to discuss the delicate dance military officers have to perform with their elected and appointed civilian leadership. Admiral Lee described the Pinnacle discussion that most resonated with him that he said he would never forget. A White House official told the small group of three- and four-star military officers, "Gentlemen, you need to be thinking about what issue, what principle, you would be willing to throw your stars on the table for, because it may come to that." Admiral Lee then stated what the White House official, "didn't say, but what he should have said is, are you willing? . . . Are you willing to step into the gap for us and for your principles?"[3]

With regard to the COVID-19 vaccine mandate for the military, the issue was not just an apparent lack of moral courage at the general and flag officer ranks. I did not know of a single active-duty general or admiral who personally refused the COVID-19 vaccination or who publicly came out in opposition to the COVID-19 military vaccine mandate. What was worse was that there were leaders at the three- and four-star ranks who were aggressively sacrificing service members to further the administration's political aims at the expense of military readiness.

The original vaccination mandate from the secretary of defense was a lawful order because he required vaccination with a fully FDA-licensed product *only*. Within my chain of command, I could clearly see that the secretary of the Navy and the chief of naval operations, four-star admiral Michael Gilday, had given similar lawful orders requiring mandatory vaccination with the fully FDA-licensed product and permitting voluntary vaccination with any EUA products. Yet, somehow, the Navy and the rest of the DoD was implementing the lawful vaccination mandate in an unlawful manner. There were no fully FDA-licensed products available, leaving only EUA products for service members to fulfil the COVID-19 vaccination requirement. The interchangeability falsehood was in full effect and unlawfully used as the justification for forcing service members to either be vaccinated with the available EUA product or be separated from the service. Thousands of service members had already been kicked out unlawfully. I began to search in earnest for the origination of the unlawful orders, since it had not begun with the secretary of defense, the secretary of the Navy, or the chief of naval operations.

I found my answer in a vaccination order sent to the force by email that I had received about two months prior. As far as I could tell, it was Admiral Christopher Grady, then in command of United States Fleet Forces Command, who had issued the first unlawful order. He was the same leader who had apparently been so eager to issue his own vaccination guidance that his mandatory vaccination order went out to the fleet days before the same orders were issued by his immediate superiors, Admiral Michael Gilday, and Secretary of the Navy Carlos Del Toro. Unbelievably, the September 8, 2021, email I received included not just the order from Admiral Grady, but an acknowledgement that there were significant concerns over the differences between the EUA product and the fully FDA-licensed product.

The email from Admiral Grady's deputy commander noted that there were discussions the previous week "about the potential for OPNAV to issue additional guidance to streamline the conversation regarding the BioNTech and Comirnaty versions of the vaccine (WRT [with regard to] EUA vs. federal licensing)." In that same email, the deputy commander attached the first interchangeability memo from the Navy surgeon general, Rear Admiral Gillingham, and stated, "Attached is a memo from the SG [Surgeon General] commemorating the facts and establishing that there is no difference amongst the two. No further guidance is anticipated at this point." Essentially, Admiral Grady was ordering vaccination regardless of the legal status of the vaccine. This was illegal, particularly since severe administrative punishment followed for service members who exercised their right to decline receipt of the EUA products.

As I was mulling over how to articulate the various violations of law and UCMJ articles in the Navy Regulation 1150 complaint I planned to file against Admiral Grady, my eyes came across the language of UCMJ Article 94—Mutiny or Sedition. During that time, I kept a large poster of all punitive articles of the UCMJ above my desk. I had placed it there in part because, as the second in command of my unit, I was responsible for enforcing good order and discipline, including adherence to the UCMJ. I had a fantastic unit and had not needed the visual very often until the military vaccine mandate, at which point it certainly came in handy. As I researched the required elements necessary to prosecute someone for mutiny, I was stunned by how perfectly Admiral Grady's actions seem to fit.

UCMJ Article 94—Mutiny or sedition, section (a)(1), states that any person subject to this chapter who "with intent to usurp or override lawful military authority, refuses, in concert with any other person, to obey orders or otherwise do his duty or creates any violence or disturbance is guilty of mutiny." To simplify this language, a service member can be prosecuted for mutiny if:

1. they refuse to obey lawful orders,
2. they do so in concert with one or more other persons, and
3. they do so with the intent to usurp lawful military authority.

In reviewing the evidence I had of Admiral Grady's actions, it appeared to me that Admiral Grady:

1. Refused to obey the law and the orders of the secretary of defense, the secretary of the Navy, and the chief of naval operations who, as required by law, ordered that only fully FDA licensed vaccine products would be used to fulfill the COVID-19 vaccine mandate.
2. Refused to obey the law and his superiors' orders in concert with the Navy surgeon general, Rear Admiral Gillingham, who issued a memorandum that substantially misrepresented the understanding of "interchangeability" to falsely convince service members that their legal right to refuse an EUA product was somehow stripped by being similar enough to the fully FDA-licensed product.
3. Attempted to usurp presidential authority by waiving EUA product informed-consent requirements that can only be done by the president of the United States, who is the ultimate military authority, and the only one who has the statutory authority to waive EUA-product informed-consent requirements in accordance with 10 USC §1107a.

When I finally completed my Article 1150 complaint against Admiral Grady, I included this explanation of the violation of UCMJ Article 94, along with discussions of several other laws that Admiral Grady appeared to violate. I submitted my complaint on November 27, 2021 to Admiral Michael Gilday, who was the general court-martial convening authority over Admiral Grady at the time. While I had decided not to fear the consequences of upholding my oath to the Constitution and opposing the apparent unlawful actions of my superiors, I was certainly concerned about being taken away from my family with no warning. I stayed up many subsequent nights, following the kids' bedtime, half expecting federal authorities to be knocking on my door, demanding I come with them under some false pretense.

Ultimately, no one came after me and my complaint was mostly ignored. My efforts seemed to have had no impact at the time as my complaint was apparently swept under the rug as quietly as possible. I found out much later that specific decisions were made to dismiss my complaint and to not place Admiral Grady under investigation. Subsequently, Admiral Grady's career progression continued unabated. Less than three weeks after my complaint against Admiral Grady on December 16, 2021, the United States Senate confirmed him as the 12th vice chairman of the Joint Chiefs of Staff.

As the vice chairman of the Joint Chiefs of Staff, Admiral Grady became the second-highest-ranking military officer in the entire US military. The highest-ranking military officer in our nation, General Mark Milley, became Admiral Grady's direct supervisor, and is the only officer in the US military to outrank him. I found it interesting that Admiral Grady was selected to become the vice chairman to General Milley. It has been reported that in the summer of 2020 General Milley wrote a resignation letter to then Commander-in-Chief, President Trump, in which he accused him of "using the military to create fear in the minds of people, being a racist, and ruining the international order."[4] General Milley decided not to resign and never sent the letter to President Trump. His differences with the president were serious enough, however, that he allegedly told his staff he would fight the president from the inside, saying "if they want to court-martial me, or put me in prison, have at it, but I will fight from the inside."[5]

This was not the only controversial thing that General Milley did while serving as President Trump's chairman of the Joint Chiefs of Staff. Allegedly fearful of potential actions President Trump might take following the 2020 election, General Milley made multiple phone calls to his Chinese counterpart in the People's Liberation Army to assure him that the US government was stable and not a threat. According to a book by Bob Woodward and Robert Costa, he told the Chinese military officer, "General Li, I want to assure you that the American government is stable and everything is going to be OK. We are not going to attack or conduct any kinetic operations against you."[6] The book also reports that General Milley said, "If we're going to attack, I'm going to call you ahead of time. It's not going to be a surprise."[7]

General Milley has since publicly defended his phone call with the Chinese official as appropriate under the circumstances.[8] He also testified before Congress that he had informed numerous Trump administration officials about the phone calls, including Secretary of State Mike Pompeo, White House Chief of Staff Mark Meadows, and Active Secretary of Defense Chris Miller.[9] I have no authority or specific personal involvement in the situation to make any judgment about General Milley's actions. As a private American citizen, however, and one who has studied American military history, I cannot imagine any point in our past when such a communication would have been an appropriate action. Giving the enemy a heads-up to ensure that any attack was "not going to be a surprise" could cost Americans dearly in blood that might otherwise not be shed.

With the American Revolution still raging in 1779, Benedict Arnold began to doubt the American experiment, believing that the attempt at independence would ultimately fail. He turned his attention inward and began making plans to maximize his own prospects. He opened secret channels with the British and began feeding them information.[10] When it became apparent he would obtain command of American forces at West Point, he offered to surrender West Point for the price of £20,000 (roughly $5 million today). On August 3, 1780, General Washington issued written orders to Major General Arnold directing him to "proceed to West Point and take command of the Post . . . [and] endeavor to obtain every intelligence of the enemy's motions."[11]

After taking command, Benedict Arnold responded to General Washington with an August 6, 1780, letter in which he began laying out the groundwork for the surrender he was secretly planning. His letter contains a litany of excuses such as a lack of fresh provisions and having to lead militia instead of regular troops. Arnold offers General Washington no solution or even a plan to overcome these challenges. In an ironic twist of historical parallels, the most significant excuse General Arnold made for his betrayal was the potential that his troops could be exposed to illness. In a letter to General Washington dated one month before the plot to betray his country was exposed, Benedict Arnold stated, "I am very apprehensive, that the troops by being exposed will become sickly, and desertions follow in consequence."[12]

I do not know if our leaders have lost faith in the American experiment, but it has been said that history has an odd way of repeating itself. It is my hope, however, that if there is a nefarious force at work against our nation, we will be able to expose it in enough time to protect our constitutional republic and our way of life. Benedict Arnold's plot was ultimately exposed before it could be executed. At the time, the American colonies were in a very precarious situation, both politically and militarily. The loss of West Point could have been a significant factor in changing the course of history. Instead, the war ended with the Treaty of Paris in 1783, putting our Founding Fathers on the path to writing and ratifying the Constitution we now swear to support and defend. Let us hope that we too can right the ship and return to following the law and protecting the constitutional rights of all Americans before it is too late.

CHAPTER 11

Unconventional Lawfare

Annihilate the profession of the law, and the liberties of the country would soon share the same fate. . . . Let them practice justice, and consider the maxim, "that can never be politically right, which is morally wrong."[1]

—Abigail Adams

I first heard the phrase "unconventional lawfare" from a thirty-year Navy SEAL master chief that I met through our growing support network. The phrase was created to be a clever play on the Joint Warfighting concept, unconventional warfare, which utilizes underground forces to disrupt a superior occupying force in a denied environment and typically over a long timeline. While the analogy between unconventional warfare and unconventional lawfare is not a perfect one, there were several parallels between unconventional military operations of the past and the current environment we found ourselves in.

In our case, the superior power is operating in an unlawful manner, denying service members' right to serve with religious convictions, and denying the right to refuse emergency use products. Even if we could avoid the impending administrative separations, we knew that a quick total victory was not likely to happen. We therefore prepared ourselves for a grueling fight over a much longer time horizon. The master chief and I discussed various options to oppose the overwhelming bureaucratic force arrayed against us that was attempting to deny us our constitutional rights. The master chief's wisdom and foresight helped shape my own mindset about the strategic problems we would likely face.

He was also deeply concerned about what our military might look like in twenty years. His own son wanted to follow in his footsteps and join the SEAL community. However, under the DoD's current COVID-19 vaccination requirement and its refusal to grant religious-based exemptions, his son would be unable to serve unless he compromised his convictions. The DoD had already begun approving a number of administrative and medical-based exemptions to the COVID-19 vaccination requirement. Those who received one of these exemptions were permitted to continue serving. The fact that the DoD was not only disapproving *religious*-based exemption requests, but were then trying to actively separate these religious service members from the military, appeared to be part of a more nefarious strategic plan. In twenty years, would there be any military leaders left with the moral courage to oppose future unlawful DoD operations or programs?

Most of us also began to admit that our many attempts to handle the unlawfulness internally had failed. The numerous communications, reports, and complaints had fallen on deaf ears. What was worse, a significant number of service members who were attempting to handle these issues internally were beginning to be met with retaliation, retribution, and disciplinary actions. The next logical place to find relief was in the US Court system. Service members began approaching lawyers and law firms they thought might help. A great many stepped up, and lawsuits replete with US military members as plaintiffs were filed in courts across the country. One of the first military lawsuits, and one that would eventually lead to a series of crucial wins in the "unconventional lawfare" operation, was filed by First Liberty Institute on behalf of twenty-six Navy SEALs, five Navy special warfare combatant craft crewmen (SWCCs), one Navy explosive ordnance disposal (EOD) technician, and three Navy divers. The lawsuit led by First Liberty Institute General Counsel Mike Berry, was filed on November 9, 2021, in the United States District Court, Northern District of Texas.[2] Holding true to their elite warrior ethos, these thirty-five Navy special operators were the first into the breach to fulfill their oaths to the Constitution. Their courage in standing up for their beliefs and their foresight in realizing early on that the fight had to be taken to the courts are a credit to them and to their warfighting communities.

As for the Navy SEAL community, their discriminatory actions played a large role in driving their unvaccinated SEALs to the federal courts to seek justice. Navy Special Warfare Command issued *Trident Order*

#12 —Mandatory Vaccination for COVID-19 on September 24, 2021.[3] This order, specifically targeting service members with religious beliefs, directed adverse treatment against special operations personnel who sought religious exemptions from the COVID-19 vaccination requirement. The order stated that "Special Operations Designated Personnel (SEAL and SWCC) refusing to receive recommended vaccines based solely on personal or religious beliefs will still be medically disqualified [i.e., non-deployable]." In a bizarre act of religious discrimination, special operations personnel who received medical accommodations from the vaccine requirement, and therefore remained unvaccinated, were *not* deemed medically disqualified and could therefore still deploy. Essentially, the Navy Special Warfare Command was publicly acknowledging that they approved of having unvaccinated personnel out on missions, as long as their vaccination status was not due to religious convictions.

There were a number of discussions during this period about how best to fight the many DoD unlawful actions, including those of Navy Special Warfare Command, within the federal court system. There were two primary legal avenues to pursue. The first was the denial of service members' First Amendment rights to the free exercise of religion. The second was the denial of service members' rights to refuse EUA products. The general consensus among the lawyers was that the First Amendment issue was likely to be the quicker way to obtain relief from the courts. My Article 1150 complaint against Admiral Grady had mainly addressed the unlawfulness of denying the right to refuse administration of an EUA product. With that complaint recently filed, I turned my attention back to the Navy's disapproval SOP and began working on a new Article 1150 complaint against Vice Admiral Nowell for denying First Amendment religious freedom rights.

Admiral Nowell had used the disapproval standard operating procedure to deny my own religious accommodation request. He signed my denial on November 23, 2021, with a signature timestamped at 12:58:47 Eastern Standard Time. I do not know exactly when the Accommodations Review Team staff member prepared and sent this disapproval memorandum to Admiral Nowell for his signature, but according to the disapproval SOP, this step would have been completed before any member of Admiral Nowell's Accommodation Review Team was required to review the details of my request. Only *after* my disapproval was on the admiral's desk awaiting signature was an Accommodation Review Team member then directed to open my religious accommodation request and begin copying details from it into a spreadsheet.

This spreadsheet—again, built *after* sending my pre-prepared disapproval memo to Admiral Nowell to sign—was designed by the Accommodation Review Team to be an important piece of evidence, as the disapproval SOP describes the spreadsheet-building step as the most critical in the process. It also notes that Admiral Nowell and his boss, Admiral Gilday, were relying on this step to be thorough and accurate. I could not initially understand why this step would be so critical to them since the review would have no bearing on the disapprovals already awaiting signature. I then realized that the only reason it mattered to them was because both the law and military regulations required a case-by-case review of the full range of facts and circumstances relevant to each request. The spreadsheet built by the Accommodation Review Team was indeed evidence, but not evidence that the Navy was following the law. On the contrary, this spreadsheet, as well as the entire disapproval SOP, was evidence that the process was a fraud, allowing the Navy to deny all religious accommodation requests while falsely claiming that a case-by-case review was done on each request.

The Navy was not unique in developing an automated system to deny religious accommodation requests. The United States Coast Guard developed a digital tool known as the Religious Accommodation Appeal Generator (RAAG) that automated the process of generating denials of religious accommodation requests.[4] The Coast Guard's RAAG system would generate pre-determined responses based on corresponding details in a religious accommodation request. The system would then export the pre-determined responses to a disapproval template explaining the disapproval to the original requester. The Coast Guard, like the Navy, maintained a near-perfect record of denials of religious accommodation requests for the COVID-19 vaccine. The automated systems of both the Navy and the Coast Guard, used to deny nearly every religious accommodation request, violated service members' constitutional rights and circumvented the review required by law and military regulation.

Not to be outdone, the Air Force also engaged in a program of systematic discrimination against service members whose religious beliefs precluded them from receiving a COVID-19 vaccination. The Air Force's discrimination likely started at the very highest levels. Secretary of the Air Force Frank Kendell apparently directed commanders to refuse to grant religious accommodations for COVID-19 vaccines.[5] The evidence, provided in court several months later, demonstrated the animosity the Air Force had towards service members with deeply held religious convictions, that resulted in real harm directed to a particular segment of the Air Force population. The Air Force had only approved eight out of 12,623 religious accommodation requests, but exemption requests that were not based on religious beliefs received the opposite treatment with a total of 1,513 medical exemption approvals and 2,314 administrative exemption approvals.[6]

The military clearly conspired to trample constitutional rights and deny the free exercise of religion to service members, which necessitated smart and capable legal action to correct these wrongs.

The First Liberty Institute case, filed in November 2021 on behalf of the thirty-five special operators, was proceeding at the speed of law, meaning relatively slowly. First Liberty Institute General Counsel, Mike Berry had made a smart strategic move by bringing in the small litigation firm Hacker Stephens, to help litigate the case. Although the Hacker Stephens firm is only a two-person firm, they brought significant constitutional and civil rights experience with a wide range of cases, including cases before the Supreme Court. As with most lawsuits against the federal government, however, non-profit organizations like First Liberty Institute, and firms like Hacker Stephens were fighting a David-and-Goliath battle against a powerful opponent that could use our tax dollars to bring virtually unlimited resources to bear. The size and power differences did not matter to the First Liberty Institute team as they slugged it out with the Department of Justice and DoD litigators through November and December of 2021. By December 22, 2021, there were a total of sixty-one different filings in the case, not including the many additional appendices and exhibits.

Meanwhile, I had finally finished the new Article 1150 complaint against Admiral Nowell for religious discrimination and was preparing to submit it, as required by regulation, to the General Court-Martial Convening Authority over him. Unfortunately, that was Admiral Gilday, who was also using the same disapproval SOP to deny any appeals of Admiral Nowell's disapprovals. Navy regulations governing Article 1150 complaints did not permit me to file the complaint to anyone else, leaving me very little hope that Admiral Gilday would take any action to correct the unlawful actions occurring within his organization. I felt a bit like someone appealing to a mob boss to get one of his enforcers to stop beating me, when that same mob boss had ordered the beatings.

I decided to inform Admiral Gilday of my concerns up front, writing that "I have deep concerns that this complaint, detailing the discriminatory disapproval process for religious accommodations in the Navy, will not be properly addressed and will instead be ignored and dismissed." In an effort to add teeth to these concerns I also noted that "I intend to copy this communication to both the House and Senate Armed Services Committees in the hope that this will ensure that all unlawful religious discrimination in the Navy is properly addressed." I submitted the Article 1150 complaint against Admiral Nowell on December 23, 2021. I did not immediately send the complaint or any related communication to either the House or Senate Armed Services Committees. At the time, I naively hoped that informing Admiral Gilday of my intent to send the complaint to the House and Senate Armed Services Committees would be enough for them to take some positive corrective action.

Instead of sending the complaint to the House and Senate Armed Services Committee, I decided I would hold that card for later. However, I did not want the complaint to languish solely within the Navy's administrative machine, benefitting no one. So, in addition to an official internal filing, I elected to share it with our growing support network. The complaint caught the eye of a Navy SEAL plaintiff in the First Liberty Institute case. He immediately passed it on to one of his lawyers, Andrew Stephens, who filed it in court that very afternoon. The next day was Christmas Eve, and the DOJ lawyers either missed the filing or did not understand its significance. They made no motion to exclude the evidence, and it was reviewed by Judge Reed O'Connor along with everything else amassed in the case over the previous two months.

Only five business days later, Judge O'Connor came back with a ruling, granting a preliminary injunction to the thirty-five Navy special operators. He also ordered the Navy to halt adverse actions against the thirty-five plaintiffs, and enjoined the Navy from applying various vaccine orders including Trident Order #12.[7] Judge O'Connor referenced the disapproval standard operating procedure in a number of places throughout his ruling and acknowledged that the "Navy provides a religious accommodation process, but by all accounts, it is theater."[8] He went significantly further in his ruling by declaring that "The crisis of conscience imposed by the mandate is itself an irreparable harm."[9]

Shortly after this ruling, the DOJ, on behalf of the DoD and the Navy, requested a partial stay in order to retain the ability to use vaccination status to determine assignments and deployability of sailors. Relying on a declaration by Admiral Lescher, dated January 19, 2022, the Supreme Court eventually granted that stay and the Navy responded by removing unvaccinated service members from operational units and denying these service members the ability to deploy in support of global US missions. Based on the incredibly low risk that COVID-19 infections were to the military aged population, this move did not seem to make much sense. The move significantly reduced military readiness for what was essentially a preventative measure of questionable effectiveness. The entire stunt appeared politically motivated.

The apparent political motivations and careerism behind these legal maneuverings became apparent to lower ranking members of the JAG community at the annual Military Law Symposium which happened to be held a day after the Supreme Court issued their partial stay. Vice Admiral Darse Crandall gave a speech to the 200 JAG officers in attendance and then took questions. From a JAG lawyer present at that event, we received the following report:

> At the close of his hour-long speech, Vice Admiral Crandall looked around the room at the audience, paused for a few seconds before saying "I probably shouldn't mention this. I hope no one is recording." He then launched into a celebration of the recent Supreme Court decision which allowed the Navy to basically reduce their own readiness by declaring unvaccinated service members to be non-deployable. Admiral Crandall bragged about the exceptional support JAG lawyers had provided to Department of Justice attorneys in defending vaccine mandates and getting the Supreme Court to validate the Navy's position that judges should "stay out" of military decision making. He then proceeded to denigrate a Navy Destroyer commanding officer who had "sidelined a billion dollar warship" because he refused to follow the mandate. Admiral Crandall publicly ridiculed the Destroyer CO to the room full of lawyers for jeopardizing his own life, the lives of his crew, and national security for personal reasons—refusing the "safe and effective" vaccine.

It was later discovered that Admiral Lescher had significant help in generating the declaration that resulted in the Navy removing unvaccinated service members from operational units and declaring them non-deployable. He admitted under oath that he received a draft of the declaration from his JAG lawyer, Captain Elizabeth Josephson, and he speculated, also under oath, that Vice Admiral Crandall likely "participated" in drafting the declaration.[10] All of this appeared to be in support of a political agenda regardless of the lawfulness of the underlying policy. Those most responsible for supporting service members' constitutional rights failed in their duty and ultimately forced service members to face the terrible choice to either violate their most deeply held religious convictions or endure significant punishments.

This crisis of conscience was brought on by military leaders chasing the prevailing winds of politics rather than their oaths to the Constitution. Some leaders did so in full knowledge of their actions. Many others simply lacked the moral courage to take a stand for what is right. The First Liberty Institute case, *Navy SEALs 1-26 v. Austin*, was the first legal win in our campaign to stop the politicization of the military that led to such unlawfulness.

I am grateful that, by the Grace of God, I was permitted to play some small role in what we hope is the first step in restoring the rule of law to the military. When writing the Article 1150 complaint against Admiral Nowell, I had not known how impactful it would be. I simply hoped that other service members, forced to endure unconscionable religious discrimination, would find hope in it and evidence to use in complaints of their own. However, because of the actions of others, including a talented legal team, a network of service members sharing resources, and a group of special operators courageously taking the fight to the enemy, the complaint against Admiral Nowell, became so much more.

If we are able to remain in the military, our efforts will be focused on restoring the Constitution to its rightful place while ensuring no leader ever again forgets the words of our great matriarch, Abigail Adams, who wrote, "[it] can never be politically right, [that] which is morally wrong."

CHAPTER 12

The Test Passers

A people fired . . . with love of their country and of liberty, a zeal for the public good, and a noble emulation of glory, will not be disheartened or dispirited by a succession of unfortunate events. But like them, may we learn by defeat the power of becoming invincible.[1]
—Abigail Adams

From the December 2020 roll out of the COVID-19 vaccines, until September 7, 2021, the DoD had no standardized policy for testing unvaccinated service members. That changed when the DoD shifted from testing based on COVID-19 symptoms to mandatory testing based on whether a service member had been vaccinated or had been granted an exemption. The September 7, 2021, DoD Force Health Protection Guidance (FHPG) (Supplement 23) clearly tied mandatory testing to the receipt of an exemption, stating that "Service members who receive an exemption from the vaccination requirement are subject to COVID-19 screening testing."[2] Based on this policy, the use of testing shifted from being an individualized health protection measure to being a coercive tool meant to harass service members who would not, or could not, become vaccinated.

In September and October of 2021, the various military services began to set COVID-19 vaccination deadlines for service members. The DoD took this as an opportunity to again shift the applicability of the mandatory testing program. Revision 1 of the FHPG-Supplement 23 tied the beginning of the mandatory testing requirement to the service member's deadline to be vaccinated for COVID-19. The new guidance stated that "Once the applicable mandatory vaccination date has passed, COVID-19 screening testing . . . is required at least weekly for Service members who are not fully vaccinated, including those who have an exemption request under review, or who are exempted from COVID-19 vaccination and are entering a DoD facility."[3]

Basic logic makes it clear that this policy was used as a coercive measure and not as a tool to protect the health of the force. Throughout 2020 almost no service members were vaccinated because the vaccines were not available until December 11, 2020. During this time there were no mandatory asymptomatic testing requirements, despite the fact that test kits were available. In 2021 as service members chose, for various reasons, to be vaccinated, that mandatory asymptomatic testing policy did not change. If the vaccine manufacturers were to be believed regarding the efficacy of the vaccines, the vaccinated should have had little concern about contracting a COVID-19 infection. For the unvaccinated, there would also be a theoretical reduction in relative risk, proportional to the percentage of the force that had been vaccinated.

When the military vaccine mandate was implemented, the vast majority of the force chose to vaccinate. The relative risk for the unvaccinated, therefore, should have been at its lowest level since the beginning of the pandemic. Yet the testing requirement was not tied to a health trigger such as the onset of symptoms, nor was the testing requirement set to some risk-related metric, such as the percentage of a service member's unit that was vaccinated. Instead, the testing requirement was tied *exclusively* to each service member's vaccination deadline date. The only logical explanation for using the testing requirement this way is that it was a tool used to coerce service members into being vaccinated or to punish them, after their deadline date had passed, for refusing the vaccine.

What most service members did not realize was that all COVID-19 testing options were, and continue to be, only approved by the FDA under Emergency Use Authorizations. The FDA has not fully licensed a single COVID-19 test kit, leaving only Emergency Use Authorized options for the DoD to utilize. As with all EUA products, American citizens never lose the right to decline their administration. Throughout the pandemic, including after the DoD mandatory testing guidance was issued, I exercised my own right to decline Emergency Use Authorized COVID-19 tests, just as I had declined to receive a COVID-19 Emergency Use Authorized COVID-19 vaccine. Until December 2021, I was not challenged about exercising my right to decline EUA testing options.

On November 24, 2021, the Navy issued their own guidance mandating COVID-19 testing for unvaccinated sailors, regardless of symptoms or personal risk. In that guidance, commanders, commanding officers, and officers in charge were also directed to "Deny entry to any service member who does not meet the requirements." Clearly, this was about compliance and not about any health protocol, because the timeline to put this order into effect was "no earlier than the active-duty deadline [to become vaccinated] of November 28, 2021 or the reserve deadline of December 28, 2021."[4] Shortly afterward, questions arose about my own choice not to submit to unnecessary testing. To leave no doubt about my intentions, I submitted a religious accommodation request to be exempted from the COVID-19 testing requirement.

My beliefs require me to apply the principles of therapeutic proportionality to all my medical decisions. This entails an assessment of the benefits of a medical intervention in light of its risks. Because it is part of my belief system, the principle of therapeutic proportionality cannot be applied only when convenient or when it doesn't cause controversy. I will not voluntarily set aside my beliefs, especially when people, programs, or policies demand that I compromise simply for the sake of compliance.

In my December 28, 2021, religious accommodation request, I reiterated to my superiors that I still had a statutory right to decline EUA products; my request was submitted only as a preemptive measure in the event that a fully FDA licensed COVID-19 testing product became available. My faith-based requirements related to therapeutic proportionality would not change if the FDA licensed a related product. If I become severely ill with COVID-19-like symptoms and needed to ensure I was receiving proper treatment, I might then conclude that a COVID-19 test was justified. However, I was not sick, nor did I have any COVID-19-like symptoms. I was also using measures such as appropriate distancing, limiting interactions with non-family, and actively monitoring myself for any and all symptoms that would be consistent with a COVID-19 infection.

In response to my December 28, 2021, religious accommodation request and my continued efforts to exercise my right to decline EUA products, my commanding officer sent me an email threatening to relieve me as executive officer. In his December 29, 2021, email, he also denied me entrance to our building and work spaces. In an odd display of mental gymnastics, he also told me that I was considered in a teleworking status, but I was not approved to telework. His exact words were, "Because you cannot access your work spaces—you are currently teleworking. However, this is not to be construed as a [unit] approved teleworking agreement." He concluded his email to me stating, "You cannot adequately nor successfully perform your duties as the Executive Officer . . . in a telework status. Your continued refusal to submit to required COVID-19 testing may result in your relief as Executive Officer." I found it illogical that my inability to perform in-person duties was being blamed on me. Throughout 2020 and 2021 I had not once tested for COVID-19, and that had never before impaired my ability to perform my duties. The unlawful policies of the Navy, the inability of leaders to think for themselves, and the lack of moral courage required of leaders to stand up for their subordinates' rights were responsible for my being unable to enter my own work spaces to do my job.

The coercive measures applied to get me to violate both my conscience and my legal rights were, unfortunately, not unique in the DoD. Hundreds of service members across the nation and overseas were dealing with similar issues, and many were threatened with far worse than being relieved of duty. Commander Liv Degenkolb had also exercised her right to decline all Emergency Use Products including test kits. She was banned from her building and left standing outside. When her one-star admiral passed her on his way into the building one morning, he stopped to speak with her. They had worked together for some time and had a professional relationship. When asked about her situation, Degenkolb informed the admiral that she was there standing up for her constitutional rights. Unbelievably, the admiral responded with, "Don't be political," before turning his back on her and leaving her outside in the elements.

Like Major General Aris and so many other senior military leaders, Liv's admiral likely understood how politics, not science or readiness, was driving the military vaccine mandate. Liv Degenkolb wanted nothing to do with the politics and was only out in the elements because she was forced to be there by her chain of command. The only other option for her would have been to follow an unlawful order, which she is obliged by her oath not to do. The admiral, having made his own political compromise, could not understand Degenkolb's uncompromising defense of the Constitution which they both swore an oath to defend. He wrongfully projected his own politicized careerism when accusing Liv of being political.

Another such situation occurred within the Navy Medical Community to Lieutenant Kristina Chang, DDS, who was banned from her building after exercising her right to refuse EUA test kits. She was threatened with unauthorized absences if she did not show up to work, so Lieutenant Chang reported for duty as scheduled. After she was denied entrance, she remained standing outside the entire day. Despite showing up for duty as required and then being forced to stand outside for multiple days, her chain of command began to mark her as absent anyway. If a sailor gets a certain number of unauthorized absences, they can be marked as a deserter and potentially serve prison time if convicted of desertion. This courageous dental officer was, of course, not a deserter. She reported for duty as required, but it was the Navy's own policies, and the egregious coercion by her superiors, that precluded her from entering the building and treating dental patients.

Her command did not stop there, however. She was told that her command was considering sending her for a psychological evaluation. When she appealed to her officer in charge, she did not get any support. Instead, the officer in charge added to the coercion by telling her that "it would be a shame if you got a dishonorable discharge and wouldn't be able to practice [dentistry] outside [in the civilian sector]." Her command began asking questions that made Chang fear for her safety. Lieutenant Chang informed me, "I believe they are using the excuse of worrying about me to conduct a psych eval. I am worried that they could force medicate me if I were to be involuntarily admitted to the psych ward." Ultimately, the Navy discharged Lieutenant Chang from the service, and listed "Unacceptable Conduct" as the reason for separation.

Prior to 2020, I would have dismissed Lieutenant Chang's psych ward concerns out of hand. However, there is evidence that this is indeed one of the new tactics being employed by those in power to silence subordinates. The case of Army Captain Seth Ritter, as detailed in a series of official Army complaints and personal audio recordings, is a prime example of this and worth briefly reviewing.[5] Over the course of several months, Captain Ritter, an officer assigned to Fort Benning, GA, noticed a number of unlawful things occurring at Fort Benning under the command of Major General Patrick Donohoe, including the altering of firearms qualification requirements to make it easier for certain female trainees.

Ritter attempted to bring these actions to his chain of command's attention but was ignored. He also felt that Major General Donahoe was acting inappropriately on social media by seeming to show favoritism and an inappropriate level of attention to young female officers and officer candidates. This mattered to Captain Ritter because Major General Donohoe was held up as the model to imitate when using social media for professional military purposes. Ritter's attempts to shine light on the various improprieties he saw resulted in swift retaliatory actions, including retaliatory investigations and a General Officer Memorandum of Reprimand.

After much prayerful discernment, Captain Ritter felt called to continue his attempts to bring Major General Donohoe's social media behavior to the Army's attention. He filed a new complaint which he hand-delivered to the Army's Criminal Investigative Division (CID). The CID did not investigate Ritter's complaint or even attempt to determine whether it had merit. Instead, after communicating with Ritter's chain of command, they held him until a member of his command came to escort him to the Behavioral Health clinic at the base hospital under the pretext of a mental health concern. Once at the hospital, Ritter was held against his will for over twelve hours, while individuals attempted to get him admitted to the psych ward.

In the course of his twelve-hour unlawful detainment, Ritter's family lost contact with him, leading to deep concerns and even panic. What his family did not know was that Captain Ritter, without any justification or connection, was being asked a series of bizarre and "politically themed questions." As Ritter lays out in his complaint, he was asked what he "thought about the 06 Jan events at the capital, if [he] believed there was voter fraud in the 2020 presidential election, and who the current president was."[6] These questions are not part of the normal battery of mental health screening questions and are so far out of bounds that they seemed a deliberate attempt not to ascertain Ritter's mental health, but to find some way to trap him or trick him into saying something they could then use to label him an "extremist." Captain Ritter has no connection to these events and had not made any public statements about them or referenced them in any way in any of his official filings about Major General Donohoe.

Calling Captain Ritter an extremist could not be further from the truth, unless attempting to defend the Constitution and uphold Army values and Army regulations have now become extremist activities. As a deeply religious Christian man with four young children, Ritter's only desire in the situation was to stand up for the Army values and for the rule of law. He was taught that it was his duty to live up to those values himself and to report any violations of those standards whenever he observed them.

Captain Ritter's family was finally able to contact both a lawyer and a civilian physician who were willing and able to get involved. At 2 a.m., they finally succeeded in getting Captain Ritter outside the hospital and to his wife and four children. His entire family had driven to the base hospital and were waiting for him in the family vehicle. They thought they were finally free but were pulled over three blocks later by military police. Seven military police (MP) vehicles eventually surrounded the family, lights flashing.

The MPs said they were sent by the command to return him to the psych ward. Ritter recorded the entire affair. I have personally listened to these recordings and heard, with my own ears, Ritter's scared children crying in the background as he and his wife were unjustifiably harassed and intimidated by a military force sent, presumably, to silence him.[7]

It was later discovered that during his detainment Ritter was prescribed a serious of anti-psychotic medications, including lorazepam and diphenhydramine hydrochloride. These were drugs that he did not need but were prescribed with the instruction to "forcibly administer (inject) to CPT Ritter if he refuses."[8] The fact sheets for these drugs note significant and serious side effects that can be exacerbated by co-use and "could have permanently altered/injured or killed him."[9] Luckily, he was able to get out of the base hospital before the medical staff could take this potentially harmful action.

While it was medical professionals and social workers in the hospital that tried to get Captain Ritter admitted to the psych ward, it was likely that command influence played a role. When the military police detained Ritter and terrified his family immediately following his hospital release, they also admitted that it was the command that had asked them to take Ritter back to the psych ward. Given that there was nothing wrong with Ritter's mind or mental state, the only logical explanation I have for the concerted effort by so many different people to institutionalize Ritter is that Major General Donahoe's command influence was pulling the strings. As far as we know there has been no investigation of this matter, despite the apparent weaponization of psychiatry against a whistleblower who should have been protected from retaliation. Had others not intervened, it's possible that Captain Seth Ritter would not have made it out of that base hospital alive.

As for me, I am very blessed that I have not had to deal with the level of retaliation and coercion that Commander Degenkolb, Lieutenant Chang, or Captain Ritter had to deal with. After being banned from the building, I continued to report to work but had to work from the parking lot. It snowed a lot that winter, so I felt bad for my team members when they had to come outside to do business with me. Ultimately, my banishment to the parking lot lasted less than two weeks. On January 7, 2022, I was removed from my position as the executive officer and ordered to report to a new command where I was told I would be exclusively teleworking for the foreseeable future. I also took my relief as executive officer as clear indication that my various complaints would also fall on deaf ears. I resolved to follow through on my stated goal of sharing my Article 1150 complaint against

Admiral Nowell with the House and Senate Armed Services Committees. I spent that evening writing the Congressional Whistleblower report about Admiral Nowell's violations, which I submitted just before midnight on January 7, 2022, the same day I was fired from my job.

Abigail Adams wrote, "May we learn by defeat, the power of becoming invincible." Those who are Christians have a special relationship with suffering, failure, and defeat. The greatest victory in history came in the course of an ignominious defeat and the unjustifiable execution of a single Man. While, not everyone who has stood up to the military's unlawful actions surrounding the COVID vaccine mandates has been Christian, the principles equally apply to those whose conscience alone compelled them to do the right thing and fulfill their oaths to the Constitution. What matters to us, in the course of these events, is not a glorious victory, but that we hold true to our faith and to our oaths to the Constitution, even if the outcome is defeat. Many of us see this time in history as a test of character more than anything else. This test, for many of us, came in the form of a COVID-19 diagnostic test we were told we had to submit to, even though it was against our consciences. We "passed" on unnecessary and punitive COVID-19 testing, and thereby passed the test of character. As something of an inside joke and a way to spread internal camaraderie, many of us have taken to calling ourselves the "Test Passers."

CHAPTER 13

The Court-Martial of Courage

I pray God, I may never be brought to the melancholy trial; but if ever I should, it would be then known how far I can reduce to practice, principles I know founded in truth.[1]

—John Adams

Military leadership across all services continued their threats of administrative punishment against service members who exercised their rights to decline Emergency Use Authorized products. Many leaders made good on these threats by taking their subordinates to Non-Judicial Punishment (NJP) hearings and awarding what punishments they could according to military regulation. NJP is unique to the military and is not the same thing as a trial by jury. NJP is held before the commanding officer and the commanding officer alone makes a ruling based on the preponderance of the evidence gathered. According to Uniform Code of Military Justice Article 15, a service member has the right to decline NJP and demand trial by court-martial. The only exception to this right was made by Congress for service members "attached to or embarked in a vessel."

One second class petty officer (E5) I spoke to was called before his commanding officer for NJP for refusing to take a COVID-19 diagnostic test when ordered. The commanding officer found him guilty and punished him by withholding half a month's pay. This petty officer felt forced into accepting the findings and punishment from the NJP because he was unable to find a JAG lawyer who would help him with his case. He experienced discrimination apparently based on his vaccination status. This junior sailor reported that the moment he told defense JAG lawyers he was unvaccinated, they told him they could not help him. Vice Admiral Crandall's command influence and apparent prioritization of politically expedient policies is where these JAG lawyers were likely taking their cues. This junior sailor was therefore punished without due process of the law and without the legal support he requested but was denied. There were many such instances of punishment throughout the military. These instances have largely been kept from public scrutiny due to the public outcry that would likely result if the American people learned how widespread the denial of constitutional rights had become for unvaccinated service members.

Many of us had discussions about how best to deal with the increasing number of NJPs we were seeing. The only real legal strategy available was to decline NJP and demand a court-martial. A court-martial would ensure that the accused service member was afforded all Fifth and Sixth Amendment rights including, amongst many other rights, the right to confront or cross examine witnesses, the right to a jury of peers, and the right to an appeal.[2] The 1977 court case, *U.S. v. Booker*, emphasized the importance of these rights, noting that any individual waiving the right to demand a court-martial "is actually waiving his rights to a full adversary criminal proceeding with its attendant Fifth and Sixth Amendment protections."[3]

The other primary benefit of a court-martial is that the details of the laws governing EUA products would get in front of a jury and into court records. A legal win at a general court-martial would set a precedent. However, as service members began declining NJPs and demanding trials by court-martial for refusing EUA products, it became obvious that military leadership was desperate to do anything possible to avoid these trials. This was especially true of senior leaders who declined EUA products. One of these senior leaders, and one of the first "test passers," was Commander Damian Kins. As the first senior naval officer fired over refusing COVID-19 EUA products,[4] Commander Kins's story is significant and demonstrates the apparent concerted effort to trample service members' rights in order to achieve policy-related goals.

Commander Kins was executing his orders to report to the destroyer, *USS Winston S. Churchill* when Secretary Austin's military vaccine mandate was issued on August 24, 2021. After reporting onboard to take over as the executive officer, Kins filed a religious accommodation request to be exempted from the COVID-19 vaccination requirement. This request was denied by Vice Admiral John Nowell on October 22, 2021. When the Navy changed its COVID-19 testing policy from a health-related effort to a program targeting unvaccinated service members, Kins prayerfully discerned that he could not participate in COVID-19 testing.

On December 1, 2021, his commanding officer wrote him orders sending him on Temporary Additional Duty to Naval Surface Squadron FOURTEEN (CNSS-14). Commander Kins's Temporary Additional Duty off-ship was possible, in part, because the ship was not operational at the time, and was unable to get underway. The ship was "out of the water in a dry dock, shafts and rudders removed, combat system removed, multiple access cuts/hull penetrations above and below the waterline, etc."[5] The ship was not even slated to be placed back into the water until April of 2022, and the crew was not scheduled to move back aboard until August of 2022.[6]

He subsequently filed a new religious accommodation request to be exempted from COVID-19 diagnostic testing requirements specifically. Without allowing his religious accommodation request to work its way through the adjudication process and without honoring his right to decline Emergency Use Authorized products, the commanding officer of USS Winston S. Churchill convened a NJP proceeding on December 10, 2021, alleging that Commander Kins had disobeyed a lawful order to test for COVID-19. Commander Kins was given less than three hours' notice of the NJP being convened. Since he had been given temporary duties off-ship, and based on the fact that the USS Winston S. Churchill was in an overhaul and was not operational, Kins attempted to exercise his right to decline NJP and demanded a trial by court-martial. A member of the JAG Corps assigned to CNSS-14, denied Kins his right to a trial by court-martial citing the UCMJ Article 15 "Vessel Exception."[7] The resulting NJP, however, could not even be held on the ship due to the extensive maintenance being conducted and was therefore held on a nearby barge. Commander Kins was found guilty of disobeying a lawful order by the commanding officer of USS Winston S. Churchill. A few minutes later, the commodore of CNSS-14 called Kins to inform him he was being relieved as executive officer of USS Winston S. Churchill and would "remain" assigned to CNSS-14 until further notice.[8]

A number of Commander Kins's rights were violated by this proceeding, including his right to consult with a lawyer, his right to receive proper notice of the proceeding, and his right to call witnesses to speak on his behalf. But most significant was the violation of Kins's right to a trial by court-martial. The vessel exception to the UCMJ Article 15 right, that a service member can decline NJP and demand a trial by court-martial, is an exception that, by legal precedent, must be very narrowly applied. Multiple courts have established that the vessel exception to an individual's Article 15 right to a trial by court-martial can only apply to "vessels at sea or about to go to sea."[9] Most significantly, the *U.S. v. Edwards* ruling in the US Court of Appeals also established that the operational status of a vessel was an important aspect in determining if a ship was a "vessel" for the purposes of UCMJ Article 15.[10]

Essentially, a ship that is not in an operational status, or not able to conduct operations for whatever reason, cannot be considered a "vessel" for the purposes of UCMJ Article 15. Therefore, any accused sailor attached to such a unit is free to demand a trial by court-martial and has an absolute right to do so. These court rulings are legally binding and have established legal precedents that must be followed by all parties, especially by the United States Navy. A violation of the legal precedent established by these rulings is a violation of the law. The US Navy, for whom the vessel exception "carve-out" of individual constitutional rights was created, has a special duty to ensure that this "carve-out" is used sparingly and only when absolutely necessary to ensure that the operations of sea-going vessels are not halted for every minor disciplinary infraction. Abuse of the vessel exception, and even the use of the vessel exception when not absolutely necessary, is an abuse of individual constitutional rights. To unlawfully apply the vessel exception as a pretext for denying these substantial rights is an egregious constitutional violation that demands rectification. But instead of correcting its mistake, the Navy accelerated efforts to remove Commander Kins from the Navy. An administrative separation process was initiated based solely on the results of his wrongfully conducted NJP.

It also appeared that the Navy was intentionally using Commander Kins's situation to make an example of him for others who might decide to stand up for their rights. Kins had intentionally refrained from discussing his situation publicly. He was operating in good faith and trusted his senior military leadership. The Navy betrayed that trust by unlawfully depriving him of his right to a trial by court-martial. They further destroyed any remaining good faith by leaking his personal situation to the media. While it is normal for military departments to report the relief of senior military officers to the press, it is not normal, nor is it legal, to also report that officer's private medical information that should be protected by the Health Insurance Portability and Accountability Act (HIPAA).

The Navy ensured that the public knew Commander Kins's vaccination status when reporting that he had been removed from his position. Dozens of news organizations picked up the story, including Kins's vaccination status and the fact that he declined COVID-19 testing. His story also went viral on social media in a number of military circles. The Navy leak apparently served its purpose because Kins later received a report from a high-ranking insider who told him that a number of senior leaders who were not intending to receive the COVID-19 vaccine succumbed to the manufactured pressure after hearing of the NJP punishment and subsequent firing. The Navy apparently punished Kins publicly—and illegally—in order to strike fear into the hearts of others and coerce them into receiving the COVID-19 vaccine.

The Navy was not the only service attempting to avoid trials by court-martial at all costs. A number of service members from other branches of service attempted to decline NJP and demand trial by court-martial. In these situations, the other services would often schedule the court-martial, but when the service member did not back down, they would cancel it and remove any NJP punishments already awarded. They then often moved straight to an administrative separation, which military regulations permit without a court-martial conviction. As much as many of us wanted to get to a court-martial to establish the unlawfulness of the military vaccinate mandate, there were proponents of the mandates who also wanted a court-martial ruling to establish that it was lawful. These individuals were apparently looking for a test case they thought would lend itself well to a legal victory. A direct victory for the lawfulness of the vaccine mandate was unlikely, so they apparently turned their sights toward mask- and test-refusal cases thinking those would be easier cases to win.

Just such a test case finally emerged, ironically, out of the Army Public Health Command based at Aberdeen Proving Ground, MD. First Lieutenant Mark Bashaw, a sixteen-year preventative medicine officer, was assigned to the US Army Public Health Center. His official duties included participating in fact-finding inquiries and investigations to determine potential risk to personnel from diseases and other non-battle-related injuries. When the pandemic manifested itself, First Lieutenant Bashaw did exactly what he was trained to do and investigated everything about the disease to ensure the Army's risk communication strategy was accurate and comprehensive.

As the COVID-19 vaccine data began revealing significant safety concerns, Bashaw attempted to communicate these risks to his chain of command. He was repeatedly shut down or ignored. For his own situation, First Lieutenant Bashaw declined the COVID-19 vaccine based on his religious convictions. He also ultimately declined both the masks and COVID-19 diagnostic tests based on their status as emergency use products. On January

12, 2022, he was charged with a violation of Article 92, Failure to Obey Order or Regulation, and was referred for trial by special court-martial. The bottom signature on First Lieutenant Bashaw's court-martial charge sheet, and the person who signed on behalf of Major General Robert L. Edmonson, was Army Colonel Yevgeny S. Vindman, Staff Judge Advocate.[11]

Army Colonel Yevgeny Vindman, now a retired lieutenant colonel, was the top lawyer at Aberdeen Proving Ground, MD and the apparent driving force behind the court-martial of First Lieutenant Mark Bashaw. He is also the twin brother of Army Lieutenant Colonel Alexander Vindman. Alexander Vindman is an interesting soldier as both a Purple Heart recipient and the key witness in the 2019 impeachment trial of President Trump.[12] In a statement that I certainly agree with, Alexander Vindman noted in his 2021 memoir that "anyone who has served would know, not following an unlawful order is also a requirement of all who serve."[13] I am not sure his twin brother, Yevgeny, would agree, however, based on his actions related to First Lieutenant Bashaw's court-martial.

In the months preceding Bashaw's court-martial, Vindman, who has a significant social media presence, issued a number of tweets proselytizing for various COVID-19 mitigation measures. On January 30, 2021, Colonel Vindman tweeted, *"As refugees from a communist dictatorship @AVindman [Alexander Vindman] and I know what oppression is. It's not a mask. Masks during a pandemic are a responsibility. Citizens have a responsibility to each other in a free republic."*[14]

On March 5, 2021, he tweeted, *"Pfizer COVID vacc #1. On my way to 95% invulnerability to COVID. @AVindman got the J&J vacc this morning. Get whatever vaccine is offered the first chance it's available to you. We must all do our part for each other and for our country. Vaccination is patriotism."*[15]

On August 7, 2021, he retweeted a picture of a baby who had supposedly been suffering from COVID-19 and had to be both intubated and airlifted 150 miles due to a purported Houston hospital bed shortage. Accompanying the retweeted picture he added the following statement: *"What a beautiful baby. She can't decide to wear a mask/get vaccinated. She must rely on adults around her to protect her and keep her safe by themselves wearing a mask/getting vaccinated. #WearAMask #GetVaccinatedNow."*[16]

On September 10, 2021 Colonel Vindman tweeted a quote from a speech given by President Biden the day before: *"POTUS: "We've been patient, but our patience is wearing thin . . . Your refusal has cost all of us." Get vaccinated!"*[17] In the September 9, 2021, speech referenced by Colonel Vindman, President Biden also made the following three statements: 1)

"The unvaccinated overcrowd our hospitals, are overrunning the emergency rooms and intensive care units, leaving no room for someone with a heart attack, or [pancreatitis], or cancer"; 2) "We're going to protect vaccinated workers from unvaccinated co-workers"; and 3) "I understand your anger at those who haven't gotten vaccinated."[18] These were the sentiments expressed by the Staff Judge Advocate over Aberdeen Proving Ground in the months leading up to First Lieutenant Mark Bashaw's court-martial at Aberdeen Proving Ground, MD.

In a show of support for Bashaw, a significant number of service members from our military network made the trip to Aberdeen Proving Ground, MD, to witness the first ever court-martial over COVID-19 related mandates. I took two days of leave to ensure I could travel and attend in person as well. Being there in person helped to ensure that we could be in the room to hear all of the arguments made regarding the lawfulness of the orders to accept EUA products. Ultimately, Mark Bashaw was found guilty because the order to be tested for COVID-19 was deemed lawful by Judge Robert Cohen. In the courtroom, I heard the judge make a significant mistake in reading and interpreting the law, however. This mistake was glaringly obvious at the time, and I was able to confirm it when I received the transcript of the trial.

To ensure that no one is confused, it is important to review the applicable EUA law, 21 USC § 360bbb-3. According to this law, EUA products come with the *Required Condition* that "individuals to whom the product is administered are informed . . . of the option to accept or refuse administration of the product." The words in the Required Conditions paragraph describing the responsibilities related to establishing conditions of authorization, state that "the Secretary . . . *shall* . . . establish such conditions as the Secretary find necessary or appropriate . . . including the following . . ." Immediately after this statement is the list of required conditions. These conditions, which include "the option to accept or refuse," are required, just like a person's Miranda rights are required. For example, if a police officer is unable to read a person their rights for whatever reason, that person does not lose those underlying rights. Similarly, if the Secretary of Health and Human Services is unable to ensure everyone is informed of all of the Required Conditions of EUA products for whatever reason, it does not mean that the required conditions are no longer applicable.

For the purposes of explaining what happened at Mark Bashaw's court-martial, it is also important to note that the word "may" is not to be found anywhere in the Required Conditions section of 21 USC § 360bbb-3.

There is a section just after the Required Conditions section called "Authority for *Additional Conditions.*" This section uses very similar language to the Required Conditions section, but instead of the word "shall," the word "may" is used. The language in the Additional Conditions sections specifically states that, "the Secretary *may* . . . establish such conditions as the Secretary finds necessary or appropriate . . . including the following . . ."

Then the law lists several additional conditions that the Secretary *may* establish for the administration of EUA products. The Additional Conditions, therefore, are the conditions that are optional based on what the Secretary may or may not choose to establish. The Required Conditions, however, are always required, and the Secretary *shall* establish them. To be clear, the option to accept or refuse EUA products is listed in the *required* conditions and not in the *additional* conditions.

Judge Robert Cohen, when verbally reading the law aloud and explaining it to the Court, superimposed the word "may" over the word "shall" in reading the Required Conditions section. He even *emphasized* the word "may" to ensure that everyone understood, incorrectly, that the word "may" was the word that applied to the Required Conditions. Judge Robert Cohen's words, quoted directly from the trial transcript, are as follows:

> It then goes into the next section talking about "appropriate conditions designed to ensure that individuals to whom the product is administered are informed." Again, this is a "may." That the Secretary authorized the emergency use of the product of any significant known or potential benefits and risks of such use, and of the extent to which such benefits and risks are unknown, and of the option to accept or refuse administration of the product.[19]

This false statement made by Judge Robert Cohen is so egregious that I am reminded of the lie told by Justice Oliver Wendell Holmes in *Buck v. Bell.* There is no way for me to know if Judge Cohen intentionally lied, but the result was similar to *Buck v. Bell* in that Cohen's ruling enabled the government to continue the violation of bodily integrity that should have had no standing in law. The falsehood established at Mark Bashaw's court-martial propped up an EUA product mandate that should have come to a swift and inevitable end on April 28, 2022, when we finally got a court to review EUA product law. Instead of protecting the rights of service members, however, Judge Robert Cohen misread the law and ensured that the current administration and politicized leaders in the military could save face with a mandate that violated the law and service members' basic constitutional rights.

Shortly after the trial, the defense realized that Colonel Vindman's politically unrestrained social media commentary could have played a significant role in the trial through undue command influence. UCMJ Article 37 deals with unlawful command influence and states, in part, that "No person subject to this chapter may . . . by any unauthorized means, influence the action of a court-martial . . . in reaching the findings or sentence in any case, or the action of any convening, approving, or reviewing authority with respect to his judicial acts."

The defense began gathering evidence of Colonel Vindman's potential undue command influence to file as a motion in court. Just nine business days after the trial, the defense was able to file a motion to recuse Colonel Vindman only to be told that Vindman had unexpectedly gone on terminal leave in order to retire from the military. This move by Colonel Vindman was surprising to me. He had only recently been promoted to colonel. Upon promotion, most senior officers stay in the military at their promotion rank for at least three years in order to retire at that rank. Colonel Vindman elected not to do this and instead chose to retire, just days after a very public trial over a very politicized issue.

Vindman's retirement intentions did not stop him from taking a victory lap at Mark Bashaw's expense, however. On April 29, 2022, Colonel Vindman tweeted, *"Proud of the prosecution team at Aberdeen Proving Ground. Secured a first in the nation conviction at court-martial of a lieutenant who failed to obey lawful orders re COVID mitigation measures. Guilty on all 3 counts."* On May 4, 2022, Colonel Vindman also retweeted an *Army Times* article about First Lieutenant Bashaw and added, *"Army officer convicted in first known COVID court-martial – Fiat Justitia Ruat Caelum [Let Justice be done though the heavens fall]."*

I found it very odd that as a senior military officer, Colonel Vindman was apparently trying to score political points by publicizing the "successful" prosecution of a junior officer under his jurisdiction. He was not a district attorney running a political campaign for reelection. I could not understand what appeared to be very questionable political speech from a military officer who had recently analyzed ways to recall retired General Michael Flynn for a court-martial over questionable political speech.[20]

Despite the incorrect ruling based on a significant falsehood, there was one fantastic positive that came out of Mark Bashaw's court-martial: the military support network finally got a chance to meet in person. On the morning of the second day of the court-martial we were also able to promote JJ McAfee to full colonel in the United States Air Force. He asked both me and Mark Bashaw to do the great honor of pinning on his well-earned eagles. The list of names of those who showed up at the court-martial reads like a who's who amongst constitutional defenders and patriots. These individuals included, First Lieutenant Olyia Bashaw, Sergeant Alexandru Julean, Sergeant First Class Joshua Snodgrass, Sergeant First Class (Retired) Corey Terry, Sergeant Avery Farson, Chief Warrant Officer 4 Brian Fitzgerald, Colonel John McAfee, Lieutenant Colonel Jon Cheek, Staff Sergeant Zarah Lacsamana, Lieutenant Colonel Rob Pike, Chief Benjamin Coker, Commander Olivia Degenkolb, Sergeant First Class Brian Stermer, Chaplain Jonathan Shour, and Major Carolyn Rocco. These individuals came from all over the nation to show their support. Major Rocco even flew, at great personal expense, all the way from Idaho to be with us.

Mark Bashaw's conviction and his courage in the face of severe injustice was a great inspiration to those of us who were there to witness it in person. We chose to take positives from the experience. We are not the type of people to stay down when we get knocked off the horse. Instead, we made use of the disappointing results at Bashaw's court-martial to steel our resolve. We also started a new practice that we called the Caravan to Freedom. Whenever there is a court hearing or a disciplinary board, we put out the word and try to fill the courtroom with uniforms. We hope that the visual of a room full of uniformed patriots will be a message of hope for those who love liberty and a reminder to all present that courage cannot be court-martialed.

CHAPTER 14

Thwarting the Destruction
of Readiness

In the presence of an enemy, to propose to take men out of the ranks . . . is to use the law not for the safety but the destruction of the nation, and merely as a cloak for delivering it up to an enemy.[1]
—Thomas Jefferson

Due to Judge Robert Cohen's false misreading of the law, our initial foray into the EUA legal battlefield was not the legal victory we had hoped for. We resolved to continue working on getting a fair hearing for our EUA legal arguments. We understood, however, that such an effort would likely take a great deal of time. Most Americans were probably not ready to hear they had been duped. They were not yet ready for the realization that EUA mandates were, and continued to be, unlawful. Over 268 million Americans have completed a COVID-19 vaccine series.[2] If only 25 percent of those did so under duress because their jobs were on the line, that would still be 65 million people. It is likely that the true number of individuals who felt coerced into getting the vaccine is greater than 25 percent. Either way, when these millions of coerced people realize, too late, that their employers and their government forced them to accept EUA products illegally, it will lead to an awakening that our country desperately needs.

As the military branches continued administratively separating service members who did not accept EUA products, a concurrent recruiting crisis hit all branches of the Armed Forces simultaneously. The Army, for example, reported falling short by 15,000 soldiers in their recruiting goals for fiscal year 2022.[3] The Navy, for their part, missed their active-duty officer, reserve officer, and reserve enlisted recruiting goals by over 2,500 recruits. The Navy barely hit their 33,400 active-duty enlisted goal, but to make their goal they poached Delayed Entry Program recruits by calling them to report earlier.[4] This move, made by multiple services, is only a temporary fix and will likely deepen the recruiting crisis in future years. To make matters worse, knowing they would be unable to make their original goals, the branches also cut their future recruiting targets.[5] Moving the goal posts to make the recruiting target easier to achieve is not going to help overall readiness. It also appears to be a move linked to political perceptions rather than war-fighting needs, designed to make recruiting appear better than it actually is. Reducing recruiting targets places political expediency above military readiness and weakens us more than simply missing the original recruiting targets.

With the recruiting crisis deepening, lawmakers began making public calls for correcting some of the DoD's recent mis-prioritizations and divisiveness in policy. Representative Mike Waltz sent a letter to US Military Academy Superintendent Lieutenant General Darryl Williams questioning the use of critical race theory materials in cadets' mandatory instruction. He noted that required lectures, including "Understanding Whiteness and White Rage," and "White Power at West Point," were part of an effort to get "future military leaders to treat their fellow officers and soldiers differently based on race and socio-economic backgrounds."[6] He concluded that he could not "think of a notion more destructive to unit cohesion and morale."[7]

The 5.3 million personnel hours spent on the *Leadership Stand-Down to Address Extremism in the Force* when an investigation revealed just 100 such cases is another example of focusing on ideology rather than on readiness. This mis-prioritization resulted in a bipartisan report from the Senate Armed Services Committee declaring that "spending additional time and resources to combat exceptionally rare instances of extremism in the military is an inappropriate use of taxpayer funds, and should be discontinued by the Department of Defense immediately."[8]

A November 2022 report from the offices of Senator Marco Rubio and Congressman Chip Roy made the argument that a number of detrimental progressive policies forced on the military were done as a left-wing social experiment and were weakening our national defense. The report, titled *Woke Warriors: How Political Ideology is Weakening America's Military*, demonstrated the harm to readiness caused by certain policies geared towards ideological change. As an example, the report points out that amongst an extensive list of historically disqualifying conditions, including peanut or gluten allergies, learning disabilities, and certain skin diseases, a new policy now permits sex reassignment surgeries for members of the military. Such a procedure can have "months-long recovery periods, with complete recovery taking up to one year for some procedures."[9] These procedures, which are now encouraged by new progressive policies, render the recipient non-deployable. Yet, these service members can continue serving with no negative career consequences.

At the same time, service members who declined to receive COVID-19 vaccines for religious reasons were declared non-deployable based on little more than the fact they had religious beliefs. Recall that Trident Order #12, as an example, explicitly made the distinction between Navy SEALs who received a medical or administrative exemption and those who applied for or received a religious exemption. The order, in a blatant display of religious discrimination, specified that personnel who refused based solely on personal or religious beliefs were non-deployable while those with medical accommodations from vaccination were still deployable. Trident Order #12 was a public acknowledgment that unvaccinated personnel can deploy and serve on missions, but not if the unvaccinated service member has religious convictions.

None of these calls to focus on readiness rather than political ideology seemed to impact the DoD's commitment to pursuing such divisive and self-defeating policies. With regard to the COVID-19 vaccine mandate, the significant concerns raised about readiness were not enough to halt the ongoing separation of unvaccinated service members from the military. By April 2022 the military branches had involuntarily separated a total of 3,400 service members for refusing to get the COVID-19 vaccine.[10] By December 2022, this number had grown to at least 8,400, but was likely higher due to the Air Force no longer reporting their separation numbers starting in July 2022.[11] Yet when confronted with the paradoxical position of a severe readiness crisis while still pursuing involuntary separations for the unvaccinated, some senior leaders were forced to admit the likely damage to readiness these separations will cause.

In sworn testimony for *Navy SEALs 1-26 v. Austin*, Vice Chief of Naval Operations, Admiral William Lescher, admitted that the Navy had a critical manning shortage of seven thousand vacant positions (billets) on Navy ships.[12] In a Navy that has around 250 active ships in commission[13] which are required to be fully manned to meet peak operational readiness standards, seven thousand vacant billets is a significant readiness concern. In that same deposition, Admiral Lescher was asked under oath about the potential self-imposed "vaccine-policy" loss of the over four thousand Navy sailors who were not vaccinated due to their religious beliefs. Admiral Lescher responded, "That would be a hard loss for the Navy . . . that would be not the best outcome for the Navy to lose that size of a Force."[14] Yet, Admiral Lescher's concern for the readiness impact that would occur from a loss of four thousand sailors did not appear to outweigh his intention to discharge those sailors from the service. Discharging service members of conscience is perpetrating significant and potentially irreparable harm to military warfighting readiness.

Since civilian and military policy makers refused to properly prioritize readiness, it was left to individual service members, several non-profit organizations, and a number of talented legal teams to stand in the gap and attempt to save the military from its own head-long rush to destruction. One of the forums we used to fight back was the at separation boards that were convened to involuntarily discharge service members. A service member who has greater than six years of service has a right to an in-person hearing by a panel of three senior members who decide the fate of the individual being referred for involuntary discharge from the military. Service members with less than six years do not get such a board, and their discharge is simply a paper-drill that takes only as long as it takes the various commands to process the discharge paperwork. For those who do get an in-person board, an individual can prepare a legal defense and appeal to the consciences and common sense of the members of the board. One of the first victories at such an administrative separation board occurred on May 20, 2022, at the separation hearing for Navy Lieutenant Billy Moseley.

Lieutenant Moseley was charged in December 2021 for violation of UCMJ Article 92—"Failure to Obey Order or Regulation" for exercising his right to decline COVID-19 EUA products. He was assigned to USS Bunker Hill which was then in a maintenance period with a significant number of systems not operational or completely removed from the ship. His commanding officer convened a nonjudicial punishment hearing, which Lieutenant Moseley attempted to decline by demanding, in writing, his right to a trial by court-martial. Like Commander Kins, whose ship was also not operational, Moseley's chain of command unlawfully denied his right to a trial by court-martial. With his Fifth and Sixth Amendment rights unlawfully denied, Moseley was found guilty at the non-judicial punishment hearing and then referred to an administrative separation board.

Lieutenant Moseley saw the Navy's unlawful actions clearly for what they were and felt his oath to the Constitution obligated him to fight back with everything he could legally muster. Armed with a legal defense grant from the non-profit organization Truth for Health Foundation, Moseley went to the civilian sector and hired one of the most talented military defense attorneys in the country, R. Davis Younts. Mr. Younts, a nineteen-year reserve JAG had also declined the COVID-19 vaccine for religious reasons, putting his own retirement on the line to stand for his convictions. He had already thrown his own significant capabilities into the military vaccine fight and had taken on a large number of clients including Navy SEALs, fighter pilots, and a number of pro bono military cases.

At Lieutenant Moseley's May 20, 2022, separation board, the panel, consisting of three Naval officers, were convinced by the evidence that the Navy's order to be vaccinated with an experimental COVID-19 vaccine was not a lawful order. They determined that Moseley had not committed misconduct by declining the COVID-19 vaccine. They also voted unanimously to retain Lieutenant Moseley in the naval service. Younts issued a press release the next day in which he detailed how "Lieutenant Moseley risked his twenty-two-year career and his military retirement because of his faith and his commitment to his oath."[15] Knowing that this legal win could not establish legal precedent, however, he also expressed hope that "this ground-breaking case sends a strong message to the Department of Defense."[16] This strong message apparently fell on deaf ears, as the DoD continued pursuing both administrative separations and unlawfully conducted denials of religious accommodation requests.

Not only did Moseley and Younts demonstrate that the military had issued an unlawful order by not making the FDA licensed product available, they also demonstrated why the military was so adamant to avoid trials by court-martial. The entire military mandate is built on incredibly shaky legal grounds. A similar outcome at a general court-martial would be a precedent-setting ruling. A number of military defense JAGs, who had not followed Admiral Crandall's lead in defending politicized policies, were convinced they could definitively win a case against the lawfulness of the military vaccine mandate if they could just get to a trial.

The Navy, in particular, was so desperate to avoid such a trial that they were willing to unlawfully use the Article 15 vessel exception to deny Fifth and Sixth Amendment rights to sailors. If senior military policy makers were half as desperate to do the right thing as they were to avoid having to admit making a mistake, we would likely not be in such an untenable situation with readiness. All the misprioritized policies, including the military vaccine mandate, amount to a betrayal of trust with service members. The current historic recruiting lows, and subsequent readiness impacts, will not improve until this trust is rebuilt.

Efforts from law firms and other non-profit organizations continued with the clear goal of expanding the few preliminary injunctions, including the one received by the US Navy SEALs in the First Liberty Institute lawsuit, into class-wide injunctions meant to protect all service members who had filed religious accommodation requests for the COVID-19 vaccine. In late February, the First Liberty Institute legal team contacted me about submitting a declaration to the US District Court in the Northern District of Texas, regarding the Article 1150 Complaint I had filed against Admiral Nowell and the subsequent congressional whistleblower report I wrote for the House and Senate Armed Services Committees on January 7, 2022.

The legal team thought that the DOJ and the DoD defendants in the *Navy SEALs 1-26 v. Austin* case could attempt to challenge the original Article 1150 complaint since my identifying information had been fully redacted when originally filed. I was happy to help and provided Attorney Heather Hacker a sworn declaration with an unredacted copy of the complaint along with the January 7, 2022, whistleblower report to the House and Senate Armed Services Committees. This new declaration was filed in court on February 28, 2022, as Exhibit 1 of the *Appendix in Support of the Plaintiffs' Motion for Classwide Preliminary Injunction.*[17]

Less than a month later, Judge Reed O'Connor issued the first class-wide COVID-19 vaccine-related preliminary injunction for military members. This ruling provided temporary protection for all service members in the United States Navy who had filed a religious accommodation request to be exempt from the COVID-19 vaccination requirement. The Navy was precluded from separating these service members while the lawsuit ran its course. In making this ruling Judge O'Connor noted that "Defendants' efforts to vaccinate the force have been overwhelmingly successful. Indeed, all 4,095 class members are part of the vanishingly small 0.6% of the Navy that remains unvaccinated. . . . Many of these class members already have natural immunity, thereby bolstering—not undermining—the Navy's goal of stemming the spread of COVID-19 and maintaining a medically fit force."[18]

Following the ruling, leaders throughout the Navy discussed the preliminary injunction as if it were a temporary halt that would be quickly dispensed with so they could get back to the work of separating us from the military. These leaders apparently did not understand what an extraordinary measure a preliminary injunction is and what it means to the likely final ruling. Judge O'Connor clearly explained this in his class-wide preliminary injunction ruling when he stated that the "preliminary injunction is an extraordinary remedy and will be granted only if the movants [the party making the motion for a ruling] carry their burden on all four requirements."[19] He then laid out exactly what these four requirements are and how the Navy Class Members met and established all four requirements. The four requirements laid out by Judge O'Connor are:

1. Establish a substantial likelihood of success on the merits of the case;
2. Establish a substantial threat of irreparable harm;
3. Establish that the balance of hardships weighs in the movants' favor; and
4. Establish that the issuance of the preliminary injunction will not disserve the public interest.[20]

The Navy class-wide preliminary injunction was just the first in a series of rulings. Attorneys Christopher Weist, Aaron Siri, and Thomas Bruns in the case *Doster v. Kendall*, requested class certification and filed a motion for a class-wide preliminary injunction for all members of the Air Force and Space Force. Judge Matthew MacFarland, in the US District Court for the Southern District of Ohio granted a class-wide preliminary injunction on July 27, 2022, for Air Force and Space Force members with sincerely held religious convictions that precluded them from receiving a COVID-19 vaccine. In his ruling, Judge MacFarland stated that "due to the systemic nature of what the Court views as violations of Airmen's Constitutional rights to practice their religions as they please, the Court is well within its bounds to extend the existing preliminary injunction to all class members."[21]

A lawsuit in the US District Court for the Middle District of Florida also emerged and eventually won a preliminary injunction for similarly situated members of the Marine Corps. The lawsuit, *Colonel Financial Management Officer v. Austin*, was filed by the non-profit organization, Liberty Counsel, led by founder Mathew Staver, with attorneys Richard Mast, Roger Gannam, and others in support. Judge Steven Merryday issued his Marine Corps class-wide preliminary injunction on August 18, 2022. In his ruling Judge Merryday went so far as to make the connection that the Marine Corps' compelling government interest might best be served by retaining the service members in the Marine Corps class action. He noted that the Marine Corps was denying religious accommodation requests without regard to the "current state of international turbulence and danger" or to the "difficulty in recruiting equivalent replacements."[22] He went on to state that "the government undoubtedly has some considerable interest in maintaining the services of skilled, experienced, highly trained, patriotic, courageous, and esteemed Marines (and service members in other branches) in whom the public has an immense financial investment and who are not typically readily replaceable."[23]

In an environment where senior military leaders seem more concerned with following political ideologies than maintaining military readiness, it has been left to others to take a stand in defense of our country and our Constitution. Individual service members, nonprofit organizations, lawyers, and now federal judges can clearly see what senior civilian and uniformed military leaders can apparently not see. Separating some of the most capable and conscientious service members in the face of growing international threats, with a concurrent and self-inflicted recruiting crisis, will have military readiness implications that could threaten the very existence of our nation. These leaders prioritized political ideologies and rigidly enforced a politically charged vaccine mandate that trampled the constitutional rights of service members. These actions destroyed the trust of service members and broke faith with the American people. The subsequent recruiting crisis is the natural consequence of such actions and will not be reversed until this unlawfulness is stopped, politicized military leaders are removed, and trust with the American people is rebuilt.

Individual service members who had the courage to take this fight to the federal courts are heroes, as are the lawyers and non-profit organizations that supported them. The federal judges who followed the law and their own consciences in taking on these cases have also courageously demonstrated what it means to be an American patriot. Many judges could have found ways to dismiss these ripe cases and avoid having to challenge a power-hungry government over a politicized vaccination mandate. The actions of these courageous individuals slowed the readiness bloodletting and stopped most of the involuntary separations long before Congress could legislate a halt to the mandate. Although the burden of maintaining military readiness should not have been placed so squarely on the shoulders of these various heroes, the circumstances of being forced to choose between faith and service to our country necessitated the subsequent lawsuits.

Senior military leaders may think, and some even say, that the vaccine is required for military readiness and the health of the force. These words, however, have no meaning when no related studies have been conducted to prove such an assertion. In fact, both the force health and force readiness data of the last two years appear to prove that the opposite is actually true. As Judge Merryday notes, "For the past two years, the Marines serving in the Marine Corps have ably discharged their duties. Almost all served at the onset of the pandemic and served successfully during peak jeopardy in the pandemic and before any vaccination against COVID-19 existed, almost all served during the height of the Delta variant surge, and almost all served during the Omicron variant. Nothing in the record establishes that preliminary injunctive relief for the religiously objecting Marines harms the public interest. . . . The record fails to demonstrate any meaningful increment of harm to national defense likely to result because these Marines continue to serve—as they have served—unvaccinated."[24] These marines and all similarly situated service members are fighting, not just to continue serving in the face of egregious government-sponsored religious discrimination, but they are fighting to defend the Constitution against an internal domestic threat that has apparently weaponized the military justice system to harm constitutional rights.

Defending the Constitution does not mean defending a physical piece of paper on display somewhere in a museum. Defending the Constitution does not mean blindly defending government institutions that have so often erred and trampled the rights of citizens. Defending the Constitution, at the most fundamental level, means to defend the constitutional rights of individual citizens, just as our Founding Fathers intended. While military members are especially trained to defend the Constitution, and the way of life it enables, from foreign aggression, there can be no greater civic calling for an American citizen than defending individual constitutional rights whenever possible. Where the two unlawfully diverge, American military members must choose to defend individual constitutional rights as their first and highest civic calling. Anything else puts us on the inevitable path towards totalitarianism and the subsequent deprivation of our God-given right to live free in His service.

The senior military leaders who have chosen political expediency and their own careers over this higher constitutional calling have chosen to use their authority *not* for the safety of our nation. They are even attempting to use military regulations to unlawfully remove service members when some of the most existential foreign threats in historical memory are at our door.

Thomas Jefferson warned about the dangers of just such actions, stating, "In the presence of an enemy, to propose to take men out of the ranks . . . is to use the law not for the safety but the destruction of the nation, and merely as a cloak for delivering it up to an enemy." Individual service members, non-profit organizations, members of the legal profession, and so many others are now taking a stand to prevent such a destruction. I pray that these actions are soon enough and decisive enough to effect the change required to keep our great American experiment alive.

CHAPTER 15

Continuum of Harm

These atrocious injuries have extinguished every remaining spark of affection for that parent country we once held so dear: But were it possible for us to forget and forgive them, it is not possible for you. . . . You can never confide again in those as fellow subjects, and permit them to enjoy equal freedom, to whom you know you have given such just cause of lasting enmity.[1]
—Benjamin Franklin to Rear Admiral Richard Howe,
July 30, 1776 (In response to an official offer of pardon
upon submission to the British Crown.)

This chapter was by far the most difficult to write of the entire book. The issue was not simply that the military vaccine mandate harmed people from such a broad spectrum of personal situations or that the number of individuals and families harmed by the mandate was staggeringly large. No, the challenge of this chapter was the need to select only a few representative stories that would convey the breadth of harm done while remaining concise and keeping the entire book within the context intended. While the purpose of this book may be best served by relating just a few stories, this does an injustice to the myriad stories not selected for print here. Later efforts must be made to gather and document the rest of those stories so they are not lost to future generations. We must not allow those in power to hide or censor those stories. They must see the light of day. The harm that has been done to so many military members and their families, as well as to so many American civilians, must be remedied and then never permitted again.

I've already discussed the harm done by denying the statutory right to decline EUA products. The most obvious, of course, is the rampant religious discrimination inherent in the religious accommodation request adjudication processes in the various military departments that necessitated the three class-wide preliminary injunctions to protect thousands of service members. Other harms incurred include the irreparable violations of conscience, vaccine injuries to otherwise perfectly healthy service members, unnecessary involuntary separations, financial and emotional harm to families, and a growing mental health and suicide crisis in the military.

The harm caused by the irreparable violations of conscience was brought on by the military's threat of adverse actions against those who had a conscientious objection to the COVID-19 vaccine. The Supreme Court has ruled that, "The loss of First Amendment freedoms, for even minimal periods of time, unquestionably constitutes irreparable injury."[2] It must be understood that simply being given the governmental ultimatum to violate one's conscience under the threat of disciplinary action constitutes the loss of religious free exercise.

When service members expressed their deeply held religious convictions that precluded them from being vaccinated against COVID-19, the DoD should have supported those service members. The DoD should have prioritized finding any possible way to accommodate the religious convictions of those service members. The law, in accordance with the Religious Freedom Restoration Act, requires such action. Instead, the DoD ignored these requirements under law, denied the free exercise of religion in the military, and doubled down with threats of severe disciplinary action against service members. These actions put service members "to the choice." As Judge Merryday stated in his August 18, 2022, ruling, "Because the Marine Corps has ordered many Marines and will order more to receive vaccination—an order that puts each plaintiff to the choice of either betraying a sincerely held religious belief or facing a substantial threat of serious discipline—the class suffers irreparable harm."[3]

Judge Merryday further points out that, as determined in *Steffel v. Thompson*, "It is not necessary that [a] petitioner first expose himself to actual arrest or prosecution to be entitled to challenge a statute that he claims deters the exercise of his constitutional rights."[4] The DoD's threats to service members constitute irreparable harm regardless of whether these threats are actualized. The Judge's own words confirm that "a RFRA objector's claim ripens—and subjects the RFRA objector to irreparable harm—upon the threat of serious discipline, not upon the realization of the threat."[5] The injunctions, therefore, have only protected those who have *not yet* had the threat of separation fully realized. The injunctions do not, and cannot, protect those who have either *already* undergone severe disciplinary action, or who have been forced to take the vaccine despite their religious objections.

What is to be done for the service members who actually got the vaccine against their conscience, religious beliefs, and better judgement? I personally know of hundreds of service members who were coerced and felt forced to accept products they did not want or need. About ten of my closest family and friends, whose value systems most closely mirror my own, were subject to the DoD vaccine mandate. Of these, two were ultimately forced to take the COVID-19 vaccine against their convictions and better judgement, due to the severe coercion they faced. They felt backed into a corner over the potential loss of employment and the threat of disciplinary action. Because of the threats and coercion used against them, they suffered the exact irreparable First Amendment harm that the judges in the three military class action lawsuits sought to avoid. How does a government begin rebuilding trust after inflicting such a choice on service members? I believe this question will haunt us for decades as we attempt to wrestle with the consequences of unlawful government actions during the COVID-19 pandemic.

The harms to service members who ultimately received the vaccine were not only spiritual but often physical as well. First Lieutenant John Bowes, an Air Force pilot-in-training who was sidelined for his own religious objections to the COVID-19 vaccine, played a critical leadership role in building a network of similarly situated pilots from all branches of the DoD. He also assisted this group in collecting and documenting evidence of military vaccine injuries when nearly every other military and medical organization was apparently attempting to hide this information. The results of this effort become a whistleblower memorandum to members of the House and Senate, first made publicly available by the Truth for Health Foundation on August 18, 2022.[6]

In this report, the authors share the heartbreaking stories of service members who were injured by the COVID-19 vaccine and who then reached out to members of our extended military support network for help. This memorandum detailed the injuries of dozens of service members; many of the reports included medical records, emergency room reports, reports from cardiologist visits, and personal statements from the injured. Of the many COVID-19 vaccine injuries included in the concerned DoD pilot memorandum, I have selected the following three that I think are important to share.

An Air Force master sergeant was hospitalized five days after the second dose of the Pfizer-BioNTech EUA vaccine. She experienced four strokes within hours of vaccination. Months later, this service member still could not see clearly, articulate thoughts properly, or even drive. This master sergeant had elected to pursue a religious accommodation request but was forced to attend a mandatory briefing first. At this briefing the master sergeant, along with the other potential refusers who were shuttled over to the briefing building with her, was told that the information they were choosing to believe was incorrect and that "the Russians and the Chinese have a big hold over our social media."[7] When service members at her briefing asked questions about the safety and effectiveness of the vaccine, or the availability of fully FDA-licensed vaccines, they were immediately shut down by presenters who told them their questions were "disinformation."

In a personal statement signed on January 9, 2022, this master sergeant reported crying all the way to her unit's vaccination site while dealing with her feelings about being forced to go against her conscience and receive the vaccine. In that same statement she noted that, "I could not risk a dishonorable discharge as that is equivalent to a felony on the civilian side. But I also did not want the vaccine based on my firmly held belief. Thus, I made a very difficult decision that went against my conscience; and I regretted that decision ever since."[8] This case, just one of many, exemplifies the crisis of conscience forced on service members who did not want to take the COVID-19 vaccine. The intangible harm to her conscience is irreparable in itself, but now she also has physical injuries that she may carry with her the rest of her life.

A Navy captain (O6) had to go to the emergency room for chest tightness and dizziness four days after receiving a single dose of the Johnson and Johnson COVID-19 vaccine. Tests at the hospital indicated possible blood clotting as well as a newly developed autoimmune disease that left his body unable to absorb vitamin B12. He reports swollen lymph nodes that cause him continual pain and discomfort. He will also have to take vitamin B12 shots for the rest of his life. In a memorandum for the record, this captain stated that "I reluctantly took the Johnson & Johnson COVID-19 vaccine in order to preserve my career and only source of income to support my wife and children. I now lament that decision and the effect it has had on my health."[9]

In his statement, this senior naval officer also called on "members of Congress to immediately investigate the damage forced vaccination is having on the men and women serving in our armed forces." The Navy captain's description of the vaccine mandate as "forced vaccination" is indicative of the way many military members view the vaccination campaign. The betrayal of trust for those injured by a medical treatment they neither wanted nor needed is not something that will be easily remedied.

A Marine Corps captain (O3) had to be hospitalized for chest pains four days after a single dose of the Pfizer BioNTech EUA vaccine. His cardiologist diagnosed him with likely pericarditis or pleurisy and recommended that the captain be exempted from receiving his second Pfizer BioNTech dose. The Marine Corps captain submitted a medical exemption request with this information. That medical exemption request was denied by his military medical officer. After contacting this medical officer to ask why his exemption request had been denied, the medical officer told the vaccine-injured Marine that "it was his responsibility from the Secretary of Navy and Secretary of Defense that everyone should be vaccinated unless [in] the rarest of circumstances."[10]

In a December 14, 2021, statement to the House and Senate Armed Services Committees, this Marine Corps captain stated that he knew he had a statistically zero level of risk from a COVID infection based on his age, fitness level, and a presumed previous COVID-19 infection. He told the House and Senate Armed Service Committees that he did not want to get the vaccine but "to my own shame and embarrassment, I regrettably succumbed to the pressure."[11] How will the DoD attempt to repair the spiritual, emotional, and physical injuries they have caused with this head-long rush into such a politically motivated mandate?

These three service members represent a tiny sample of the thousands who have been physically injured by the COVID-19 vaccine. Many vaccine-injured service members had religious convictions that precluded them from receiving the vaccine. The coercion and threats that pushed these service members into violating their consciences amount to a campaign of psychological warfare that has now claimed thousands of victims. Service members in this situation should have been supported by the DoD and protected by RFRA. Instead, these service members became the chief targets of the DoD's unlawful vaccination campaign.

Service members who were already separated were not protected by the preliminary injunctions, and the 2023 National Defense Authorization Act passed by Congress did nothing to remedy the harm done to them. For those already involuntarily separated, the substantial threat of serious discipline was acted upon by the military and realized by their many victims. To be readmitted to service under normal circumstances, these service members would have to apply to the Board for the Correction of Military Records, a process that often takes years. Board members would have to concur that the military had committed some wrong to justify a readmission to the military.

Under the current environment, I struggle to see this occurring. Will the military fully restore those unlawfully discharged, or will they continue to betray the trust of veterans and the American people by waiting until forced to do so by the legislature or the judiciary? Even if it is a law or court ruling that forces the DoD to take restorative action, service members should be given the choice to return or not. I find it highly likely that most would refuse to return to service, while taking only whatever back pay is owed them and remaining civilians where they would have less chance of being further harmed. If military leaders want to prove me wrong, they have a significant amount of work to do to repair the harms committed and restore the broken trust with service members and the American people.

Targeted travel restrictions directed at unvaccinated service members caused many military families severe emotional and financial hardship. We received reports of a number of service members stuck overseas or at US duty stations far from their families due to being unable to travel home unvaccinated. I think it likely that this tactic proved to be an incredibly useful tool to coerce service members who would not otherwise have been vaccinated. A number of service members stayed resolute and did not succumb to the coercion, restrictions, or threats.

Navy Musician First Class Drew Stapp was kept from his family for a total of eight months, despite having valid and approved retirement orders in hand. It took the Navy's preliminary injunction for his command to finally permit him to execute his final retirement move to his family, who had already established a homestead in Texas.

Navy Chaplain Jonathan Shour was prevented from completing his permanent change of station orders and was forced to live out of a single hotel room for nearly five months with his pregnant wife, their three children, and their family dog, hundreds of miles from his ultimate duty station. They were not provided enough funding to cover their hotel costs and were forced to pay thousands of dollars out of pocket to cover basic lodging needs.

Air Force Staff Sergeant Michael Morrisette was deployed to South Korea when the military vaccine mandate was put into place. He was scheduled to rejoin his family at his next duty assignment in June of 2022. Instead, he was held on station until November of 2022 because he was unvaccinated. The only time Staff Sergeant Morrisette was able to visit with his wife during the nearly eighteen months they were separated was during visits she made to South Korea at great personal expense. These restrictions, based on little but fear, came at great emotional and financial costs to service members and their families. When and how does the DoD start to rebuild trust with these betrayed service members?

Of all the stories I feel called to share, the story of a Navy SEAL named Daniel gives me the greatest pause. To protect his family's privacy, I will only be using his first name. It is with great trepidation that I share Daniel's story, both due to the gravity of the situation involved and due to the great sensitivity that his story demands. As a Navy SEAL, Daniel was one of the most highly trained and physically capable warriors our country had to offer. After his last deployment rotation overseas, Daniel expressed serious misgivings and frustrations about his job in the military. He communicated his intention to leave the military and his desire to move on with the rest of his life as a civilian. Before he could fulfill his goal, Secretary Austin put the military vaccine mandate into place. Daniel, already frustrated with the military, told his chain of command he would not receive the vaccine either. Although he may not have expressed it that way, he was in fact exercising a right he had, by law, to decline EUA products.

Along with other SEALs who refused the COVID-19 vaccine, Daniel was sidelined, ostracized, and isolated. His was denied the opportunity to train and continue to hone his hard-earned skills. For extended periods of time, his chain of command completely ignored him, only occasionally providing unfulfilling and menial tasks to perform. Leadership even removed the unvaccinated SEALs' badge access to their own building. They had to wait outside for building access. If permitted to enter their own building, these warriors, who had already sacrificed so much for their country, would have to endure the insult of being escorted like any untrusted foreign visitor. For nearly a year, Daniel endured this treatment, ever unsuccessful in leaving the military as he desired. Then, on September 17, 2022, Daniel was found dead by apparent suicide.

It was only a few days later that I was alerted to the situation. I was told that the commanding officer of Daniel's unit called an all-hands meeting shortly after being notified of Daniel's death. At that meeting he informed the command of the situation. He also directed those present not to speak to the media or discuss the situation outside the unit. Assuming there were no unvaccinated service members at this meeting, he also told those present not to talk to any of the unvaccinated SEALs about what had happened. Unvaccinated

SEALs were at that meeting, however, because it was an all-hands meeting and they were part of the command, albeit an unwanted part. They did not realize yet how isolated and unwanted they were by the command.

Perhaps the commanding officer did not understand, but for the last months of his life, the other unvaccinated SEALs were the closest team members Daniel had. If there were any individuals to focus compassionate leadership towards it should have been the other unvaccinated SEALs. They were with Daniel the most during those months of isolation, assigned the menial and unfulfilling jobs together. Essentially, they were Daniel's closest companions during a time of deep psychological abuse. The last official duty Daniel performed for his country before he died, was basic lawncare maintenance outside the command headquarters he was no longer permitted to enter.

Daniel was an American warrior, a highly trained Navy SEAL, and most importantly, a child of God. Daniel deserved better from the military and from his leadership. His story is one of deep tragedy and unconscionable harm. No injunction can protect him now or replace what his family has lost. He suffered irreparable harm of the most tragic kind. It particularly aggrieves me that Daniel's chain of command apparently asked for silence and a blackout of information about him and his situation. I hope I am wrong, but I can only read careerism, ambitious self-preservation, and politicization into a desire for silence. I, therefore, will do what I feel justice and my conscience requires for Daniel and any other service members who may be feeling sidelined, ostracized, and isolated. I provide Daniel's story to the American people in the hope that our nation takes action to ensure that such a tragedy can never happen again.

Sadly, the DoD has been mismanaging mental health and suicide risks for quite some time. Unfortunately, the COVID-19 pandemic gave them an opportunity to place even less emphasis on *mental* health and suicide, as they chose instead to place service members' *physical* health preservation at the very top of their priority list. The contrast in attention paid to COVID-19 infections versus suicides was not lost on anyone paying attention. Captain Courageous, the whistleblower who worked closely with Vice Admiral John Nowell, told us that he brought up the difference in COVID-19 deaths and suicide deaths to Admiral Nowell on one occasion. Admiral Nowell responded by angrily retorting that Captain Courageous was "not taking COVID seriously."

Unfortunately, it appeared the opposite was true; DoD leadership was failing to take suicides seriously. They had chosen instead to throw all their resources behind the politically expedient "threat" of COVID-19.

As discussed in Chapter 5, the DoD had even placed COVID-19 above other nation-state adversaries in the prioritization of threats. After seeing the numbers, however, it is obvious to anyone willing to follow their consciences rather than politics, that they should have placed mental health well ahead of COVID-19 in terms of readiness priorities.

Comparing the number of military suicides to military COVID-19 deaths is a shock to those who have never seen the difference before. According to the Defense Suicide Prevention Office (DSPO), in the *thirty-three months* from January 1, 2020, until September 30, 2022, there were a total of 1,460 suicides in the military.[12] In the *thirty-six months* from January 1, 2020, until December 31, 2022, despite the additional three months, there were only 96 deaths attributed to COVID-19 infections. In fact, at the time Secretary Austin mandated the COVID-19 vaccine for the military in August of 2021, there had been fewer than *thirty* military deaths attributed to COVID-19 infections. At that same time, there had already been 839 suicides since the beginning of the pandemic. The DoD should have known better and prioritized saving the lives of struggling service members. Instead, we are left with the additional harm caused by the DoD's mis-prioritization of COVID-19.

I think the continuum of harm described herein echoes the harms inflicted on colonial Americans by the tyrannical British. Armed with the power to grant pardons and amnesty, British admiral Richard Howe was sent by the British Crown to seek reconciliation with the American colonies. In July of 1776, Admiral Howe made contact with the Continental Congress and offered peace, pardon, and amnesty. The primary condition of that offer was submission to the British Crown. Congress commissioned Benjamin Franklin to respond, which he did that same day.

Citing British atrocities such as the burning of American towns in winter, and the British incitement of American Indians to massacre colonial farmers, Benjamin Franklin informed Admiral Howe that such an offer in the face of severe injuries could only increase resentment. Franklin also demonstrated a keen insight into human nature noting that the limiting factor to such a reconciliation was not the injured parties, but the perpetrators of such serious harm. He shared with Admiral Howe that even if it were possible for the American colonists to forgive and forget, "it is not possible for you (I mean the British Nation) to forgive the People you have so heavily injured; you can never confide again in those as Fellow Subjects, and permit them to enjoy equal Freedom, to whom you know you have given such just Cause of lasting Enmity."[13]

Even if those who have been injured by the military vaccine mandate forgive and forget, I do not think it is possible for those who have perpetrated this unlawfulness to return to freedom-loving ways, having so recently trampled constitutional rights and caused irreparable harm to service members and their families.

Like our colonial forebears, those who have been harmed by the unlawfully implemented military vaccine mandate are further harmed by false overtures of pardon and amnesty. We do not seek pardon for following our consciences. We do not seek pardon for defending the Constitution. In his letter, Benjamin Franklin provided the British several possible actions that could help recover the colonists' good graces and enable future potential alliances. These actions included rebuilding burnt towns and "repairing as far as possible the mischiefs done us."[14] Modern Americans injured by the vaccine mandate also require positive corrective actions to enable the rebuilding of trust.

First, those who have willfully perpetrated unlawfulness in implementing the COVID-19 vaccine mandate have made themselves enemies of the Constitution and must be held accountable. Second, the government must repair, as far as possible, all who have been harmed by the COVID-19 vaccine mandate, including the families of those whose harm is now tragically irreparable. Finally, controls must be put in place to ensure a return to the rule of law and an immutable adherence to the Constitution. Those resisting unlawful orders have never stopped defending the Constitution, while those who have perpetrated unlawfulness have stopped defending the Constitution and rule of law. If these individuals wish to be reunited with us, they must rejoin us in defending the Constitution and living up to the oath we have all taken.

CHAPTER 16

The Cover-Up

The public reports from every quarter are so strong as that no jury, on the evidence before us, ought ever to acquit . . . much less ought we to be dupes of it, whose discretion has been substituted by the law, that cases, which would forever be covered up in fraud, might not escape the punishments of the law.[1]

—Thomas Jefferson

By the early summer of 2022, the public reports of DoD unlawfulness were pouring in "from every quarter," as Thomas Jefferson might say. The extent of the harm being caused to service members and their families was beginning to unfold. The American public had also begun taking notice of the military recruiting crisis. By this time, thousands of service members had submitted hundreds of thousands of pages of internal complaints, accommodation requests, reports, and memoranda documenting the rampant unlawfulness committed by military leadership. Many service members in our extended support network, including myself, had also appealed to Congress for support.

Some service members began exercising their First Amendment rights to appeal directly to the American people. Since all internal avenues of redress had failed, perhaps if the American people were alerted to what the DoD was doing, they would demand change. I wholeheartedly supported these efforts and helped out wherever I could. However, I foresaw one last opportunity for internal action that I felt had to be attempted before all internal efforts could be declared exhausted. For me, the issue I felt still needed to be officially addressed was the unofficial "top-cover" being provided for those who were committing unlawfulness.

Normally, a service member facing such challenges, who has exhausted all other remedies, can go to the Inspector General (IG) of their service. Historically, such a service member could expect a fair and unbiased investigation if the complaint was credible. However, after the filing of hundreds, perhaps thousands, of IG complaints related to COVID-19 product mandates, we had seen nothing but dismissals. Nearly every single IG complaint we submitted was dismissed out of hand with little to no justification. Not once did I see a single response from any IG complaint that actually addressed the merits of the complaint. The response we received in nearly every single case that actually provided a justification for the dismissal was the inspector general version of the Nuremberg Shrug: "This is DoD policy."

My own complaint against Admiral Grady for illegally mandating EUA products in concert with the Navy surgeon general, Rear Admiral Gillingham, was dismissed with no justification at all. All I received was a memorandum from the Office of the Naval Inspector General stating that the matter had been evaluated and dismissed. Ironically, that memorandum was signed on December 22, 2021, just four days after Admiral Grady was confirmed by the Senate as the 12th vice chairman of the Joint Chiefs of Staff. I did not receive a response from my complaint against Admiral Nowell for religious discrimination in his use of the dismissal standard operating procedure to adjudicate all Navy religious accommodation requests.

However, as the weeks turned into months, our network began building a growing repository of small legal victories in federal court. The evidence we presented in federal courts to secure these small legal victories was first presented to the inspectors general of each service. The inspectors general ignored our evidence, and the pile of IG complaint dismissals continued to grow at an alarming clip. I began to realize that we had a severe problem with the inspectors general. They were ignoring legitimate evidence, evidence that had been significant enough to convince federal judges that violations were occurring. They were no longer defending the Constitution and the rule of law. Instead, they were defending the institution of the military, choosing political ideology and their own careers over protecting service members seeking help.

In evaluating the best way forward, I first decided to visit the Naval inspector general's website. The historical motto of the Naval inspector

general was "The Conscience of the Navy." I had not heard that the motto
had changed, yet nowhere on the website could I find "The Conscience of
the Navy." *"Apparently, to be a good cover-up artist, you must no longer have a
conscience,"* I thought. I didn't even know who the current Naval inspector
general was, so I decided to find out. I clicked on the leadership biography page and was startled to see a very familiar face staring back at me.
Admiral John Fuller, the Naval inspector general, had been my very first
commanding officer on my very first ship in the Navy. Upon reporting to
the destroyer, *USS Mason*, then-commander John Fuller was the commanding officer who greeted me.

Although Admiral Fuller and I served together for only a short time,
I believed him to be a solid leader and a good mentor to the sailors he led.
My personal experience with Admiral Fuller did not raise any red flags for
me. I had great respect for him and no reason to question his loyalty to the
Constitution or to the rule of law. However, under the circumstances and
in light of the significant evidence of unlawfulness being ignored by the
Naval inspector general, I felt compelled by my oath to document Admiral
Fuller's apparent dereliction of duty and request a full investigation.

Deciding how to file such a complaint was another challenge. I could
file another report to Congress. However, my report to the House and
Senate Armed Services Committees about Admiral Nowell's unlawful religious discrimination had not yielded any positive results. It had taken the
federal courts, and Judge Reed O'Connor, to turn the evidence of Admiral
Nowell's unlawfulness into something that would actually protect service
members. So, another appeal to Congress was probably not going to help in
this situation. I could file an inspector general complaint, but that would be
essentially asking Admiral Fuller to investigate himself. I had trusted him
once, as a young junior officer, but Admiral Fuller had destroyed all trust by
his actions and, more importantly, by his inaction. Therefore, an inspector
general complaint against the Naval inspector general himself was certainly
not the way to go.

I came to the conclusion that another Article 1150 complaint was the
best option for this situation. But an Article 1150 complaint had to be
filed within ninety days of when I learned of the wrong being done to me.
Admiral Fuller had dismissed my complaint against Admiral Grady on
December 22, 2021, much more than ninety days in the past. I had also
not received any word on my complaint against Admiral Nowell, and regulations did not permit me to file an Article 1150 complaint on behalf of
others and their wrongfully dismissed IG complaints. So, I was unable to

file a complaint based on the dismissals coming in for other members of our extended support network. I elected to keep this effort on the back burner while I continued researching.

I finally caught a break on August 5, 2022, when I received an email from the office of the Naval inspector general informing me that they had dismissed my Article 1150 religious discrimination complaint against Admiral Nowell. I was pretty shocked to read their justification. The email notification informed me that Admiral Nowell's actions "did not warrant an investigation by this office because we did not find sufficient evidence to constitute a credible allegation of misconduct by a DON senior official."[2] They had sat on my complaint for over seven months. They had not investigated anything during that time, because they just admitted not opening an investigation into the matter.

Why did this take seven months? Also, how did they "not find sufficient evidence" with all the evidence I had provided them? They knew for a fact that I had also handed this evidence over to both Congress and the federal courts. The evidence I provided was "sufficient" enough for a federal court to grant a preliminary injunction to thirty-five Navy special operators in January 2021. It was also "sufficient" enough for Judge O'Connor to expand this preliminary injunction into a class-wide injunction, protecting all Navy service members from being separated over the COVID-19 vaccine. Yet, somehow this evidence was insufficient for Admiral Fuller.

I knew the Naval inspector general would be desperate to cover-up and dismiss my complaint because he had proven to value protecting the institution over the rule of law. I was just very surprised to see such an unimaginative and false justification, when the evidence I had provided was basically doing the Naval inspector general's job for him by virtue of the federal courts. Despite Admiral Fuller's apparent obstruction, inaction, and dereliction of duty, thirty-five Navy special operators, a savvy legal team, and a judge willing to follow the law instead of politics used the evidence I provided to protect Navy service members from the harm of being unlawfully separated from the service.

Notwithstanding the false and unimaginative justification they used to dismiss my complaint, I had what I needed to push forward and expose the cover-up being orchestrated by the service-level inspectors general. I would do so with an Article 1150 complaint against Admiral John Fuller for dereliction of duty and covering up the unlawfulness of Admiral Grady, Admiral Nowell, and others. With all the other concurrent projects I was working on at the time, it took me another twenty days to complete my research and finish writing the complaint. I submitted the Article 1150 complaint against Admiral John Fuller on August 26, 2022.

This complaint was also unusual with respect to who received it, because the naval inspector general works directly for the secretary of the Navy. In this situation, the guiding instruction directed me to file my complaint to the assistant secretary of the Navy for manpower and reserve affairs. The Honorable Mr. Robert Hogue, the acting assistant secretary of the Navy for manpower and reserve affairs, was someone whose name was quite familiar to me. After Rear Admiral Gillingham had initiated the "interchangeability" misinformation campaign, Mr. Hogue had piled-on with an interchangeability memorandum of his own. In his memorandum, dated September 8, 2021, Mr. Hogue claimed that "Navy medical providers can use Pfizer-BioNtech doses previously distributed under the EUA to administer mandatory vaccinations."[3]

As extensively detailed in Chapter 8, the claim that the EUA product could be mandated based on being medically interchangeable was an incredible deception. Proclaiming interchangeability did not magically transform the EUA product, which comes with the right to refuse, into a fully licensed product. The EUA product did not go through the process required by law to be declared interchangeable with any other product. Also, by law, the process for declaring a product interchangeable with another may not occur less than twelve years after the licensing date of the original reference product. Therefore, no product could legally be declared interchangeable with any licensed COVID-19 products until at least August 24, 2033.

Mr. Hogue had helped propagate and legitimize the false claim that these products were legally interchangeable and could therefore be mandated. Yet this unlawful enforcement of the mandate was exactly the crime I was reporting. It was to Mr. Hogue that I would have to send my complaint about Admiral Fuller's dereliction of duty and cover-up of potential federal crimes. I knew the complaint would not get a fair review. I also knew that a real investigation would never occur while these individuals were now apparently being forced to cover for each other. Fair reviews and a real investigation would have to come later. Yet, at that time we still needed to speak truth to power and force them to confront their own lawlessness in a forum that required them to provide a response.

An inspector general complaint can be dismissed without being addressed. A congressional inquiry can be ignored, and very often was ignored by these leaders during the COVID-19 pandemic. The guiding instructions for Article 1150 complaints, however, required them to respond.

That is why the Naval inspector general's office was forced to notify me of the dismissal of my complaint against Admiral Nowell, when I am sure they did not want to do anything of the sort.

I did not shy away from my purpose and therefore addressed Mr. Hogue directly in my complaint. I noted that I was deeply concerned that he would be unable to investigate the complaint in an unbiased manner due to his deception regarding "interchangeability." I also said he had allowed his "memorandum to be used fraudulently to impress upon sailors that they did not have a legal right to decline administration of an EUA product." For these reasons I asked Mr. Hogue to recuse himself from ruling on the complaint I filed against Admiral Fuller. The first step in correcting a wrong is to admit that a wrong has occurred. I was hoping to at least get someone in authority to acknowledge that mistakes had been made regarding the implementation of the COVID-19 vaccine mandate.

The cover-up, particularly the cover-up of the interchangeability false-hood, was actually much worse than most people realized. It wasn't just that military leaders were forcing EUA products on Americans when the law states they have a right to refuse. It also wasn't simply a matter of demonstrating that they had executed a misinformation campaign based on "inter-changeability." The issue was that they did all of this knowingly. It was not a mistake. The most senior leaders at the top, including Admiral Fuller's boss, Secretary of the Navy Carlos Del Toro, knew that what they were doing was not legal, and their own court filings prove it.

In the lawsuit *Coker v. Austin*, filed by Attorneys Brandon Johnson, Ibrahim Reyes, and Travis Miller in the US District Court for the Northern District of Florida, the plaintiffs sought injunctive relief from the COVID-19 vaccine requirement based, in part, on the lack of a fully licensed product and attempted mandate of EUA products. The defendants in this case included Secretary of Defense Lloyd Austin, Secretary of the Navy Carlos Del Toro, and others. In responding to the plaintiffs' claims, the defense brought in an FDA expert by the name of Dr. Peter Marks to explain why an injunction would cause harm to the government's interests.

In a sworn declaration to the Court, Dr. Marks stated that an injunction "would call into question the data supporting FDA's determination that Comirnaty is safe and effective."[4] He also asserted that an injunction could "seriously undermine the government's efforts to encourage vaccination in all eligible populations by exacerbating vaccine hesitancy."[5] Dr. Marks also cited vaccine hesitancy and vaccine misinformation as one of the "most significant barriers to widespread vaccination."[6] These words, provided to the

Court on October 21, 2021, seem gravely erroneous today based on what we now know about the effectiveness of the vaccine and the alarmingly bad safety profile that has since been publicly confirmed. Also in that same testimony, Dr. Marks, likely without knowing he did so, completely destroyed the military's justification for mandating EUA products.

Dr. Marks clarified what the FDA meant by their statement that the "licensed vaccine has the same formulation as the EUA-authorized vaccine, and the products can be used interchangeably to provide the vaccination series without presenting any safety or effectiveness concerns."[7] He stated that the "FDA included this clarification in the authorization letter to avoid the unnecessary operational complications that may have resulted if pharmacies or other healthcare practitioners had believed that individuals who had received Pfizer-BioNTech for the first dose were not authorized to receive Comirnaty for the second dose, or vice versa."[8] Essentially, if a patient starts a vaccine series with the licensed version, they can finish the series with the EUA product. This is a purely *medical* determination and has nothing to do with the *legal* status of the vaccine, the rights that come with one legal status versus the other, or whether or not either version can be involuntarily administered.

In his declaration Dr. Marks immediately points out that this medical determination "should not be confused with the statutory interchangeability determination that FDA may make [under 42 U.S.C. § 262(k)]."[9] He goes on to state that "The statutory interchangeability determination requires a licensed reference product and a subsequent applicant seeking licensure, which is not present here . . . While FDA determined Comirnaty and Pfizer-BioNTech Covid-19 vaccine are medically interchangeable, there are legal distinctions between BLA-approved and EUA-authorized products."[10] One of the most critical of those legal distinctions is that every single American has the absolute right, by law, to decline the administration of EUA products. What the FDA's own Dr. Marks told the Court is exactly what we had been telling our leadership, the Inspectors General, Congress, and anyone else who would listen. EUA products are legally distinct from fully licensed products, and a mandate of EUA products cannot be legally enforced. Despite the fact that we were correct, we were ignored by nearly everyone.

Remember, this testimony from Dr. Marks was provided by the *defendants* in the *Coker v. Austin* case. Because one of the defendants in the case was Secretary of the Navy Carlos Del Toro, it is impossible to read this testimony and view the Navy's false interchangeability argument as ignorance of the law. Rather, the defendants in that case, including Admiral Fuller's own immediate supervisor, Secretary Del Toro, had positive knowledge of the law governing interchangeability as early as October 21, 2021, through the testimony of their own FDA expert, Dr. Peter Marks. The fact that they continued pushing the fraudulent narrative that the two products were legally interchangeable after October 21, 2021, demonstrates willful negligence and likely substantiates criminal activity.

The naval inspector general was given this information in various complaints from a significant number of service members. Why did he ignore it? Was he covering for Admiral Grady and his boss, Secretary of the Navy Carlos Del Toro? Secretary Del Toro knew as early as October 21, 2021, that the military vaccine mandate could not be legally enforced when only EUA products were available. Yet they continued with the vaccination campaign and severe punishments for service members who did not consent to the administration of EUA products. They even destroyed their own readiness by separating many thousands of service members as punishment for resisting these crimes.

All the while, the naval inspector general had the ability and the duty to investigate these credible allegations and get to the truth. Vice Admiral John Fuller refused to do his duty. An investigation into my complaints, and the complaints of hundreds of other service members, would have easily uncovered the facts about the availability of fully licensed COVID-19 vaccines, and the facts about the FDA's own interpretation of what they meant by the medical interchangeability of an FDA-licensed product with an EUA-authorized product. The military vaccine mandate should have ended right then, that day, on October 21, 2021, based upon the testimony of Dr. Peter Marks. Admiral Fuller had the chance to protect vulnerable service members from unlawful orders and unjust consequences. Instead, he chose to ignore his duty and cover-up these potential crimes by electing not to investigate the many complaints submitted to him.

On September 12, 2022, a whistleblower publicly released a memorandum from Acting DoD Inspector General Mr. Sean O'Donnell to Secretary of Defense Lloyd Austin titled *Denials of Religious Accommodation Requests Regarding Coronavirus Disease-2019 Vaccination Exemptions*. Mr. O'Donnell wrote this memorandum, dated June 2, 2022, to inform Secretary Austin of "potential noncompliance with standards for reviewing and documenting the denial of religious accommodation requests of Service members identified through complaints to my office."[11] The memorandum detailed a few of the various unlawful ways the services were reviewing and denying religious accommodation requests.

I saw this memo that same day, and I noted that it was an explosive confirmation of the unlawful activities that had resulted in preliminary injunctions for three different branches of the military. Similar to Benedict Arnold's manufactured excuses to General Washington, it appeared that the DoD Inspector General was attempting to manufacture a plausible excuse for himself in case later investigations wanted to know how violations of the law were proven in multiple federal courts but somehow IG complaints detailing these same activities were ignored by the DoD IG.

No news organizations were running with the story yet, so I informed the website TerminalCWO[12] of the public release of the DoD IG memorandum. TerminalCWO is a grass-roots investigative journalist outfit run by military members and veterans on a part-time basis. Concerned by the ever-growing amount of corruption they were seeing within the upper ranks of the military, the individuals who started TerminalCWO have years of experience in gathering information from service members being harmed and getting this information out to the public to prompt change. They learned long before I did that DoD leaders fear bad press more than almost anything else. The folks at TerminalCWO turned the DoD IG story around in a day and on September 13, 2022, news organizations across the country picked up their story about the admission from the DoD IG that unlawful religious accommodation review processes were occurring in the military services.

The DoD IG may have been trying to make it appear they were looking into the issue, but something did not sit right with me about Mr. O'Donnell's memorandum. I went back to the naval inspector general's August 5, 2022, dismissal of my complaint against Admiral Nowell to see if there was any mention of the DoD Inspector General. Sure enough, Admiral Fuller's dismissal notification stated that the "Department of Defense Office of Inspector General (DoD OIG) reviewed this matter and agreed with this office's conclusions. Based on DoD OIG's concurrence, this case is now closed." So, the DoD IG was telling Secretary of Defense Lloyd Austin that the services were adjudicating religious accommodation requests unlawfully, but when the actual complaint against Admiral Nowell that had started all the legal wins finally hit their desk, the DoD IG "found insufficient evidence" to investigate the matter.

The DoD IG memo to Secretary Austin was sent on June 2, 2022, and the dismissal of the complaint against Admiral Nowell was sent to me two months later. The DoD IG was basically speaking out of both sides of his mouth. My only conclusion was that Mr. O'Donnell was attempting to set up Admiral Fuller and the other uniformed Service-level Inspectors General to take the fall. He could pull out his June 2, 2022, memo at later investigations and claim he was doing everything he could. Had the whistleblower not gotten Mr. O'Donnell's memorandum out when he did, the fraud may have remained hidden.

The apparent fraud and deception were significant. It was essentially covered up by those who had the highest duty to get to the truth. The inspectors general of the services and the inspector general of the DoD failed in their duty to protect service members from criminal behavior within the highest ranks of the military. I was personally and deeply disappointed in Vice Admiral John Fuller for his apparent dereliction of duty and for allowing such serious unlawfulness to go on for so long. He was my first commanding officer and a man I had respected. Disappointment and betrayal, however, will not stop those of us who feel compelled to risk everything to defend the Constitution. Like Thomas Jefferson, we will not be the dupes of fraud. We have the law to guide how we handle the overwhelming evidence of unlawfulness before us. In the face of such rampant and public evidence, we must not allow these cases to remain forever covered up in fraud. There is a reckoning coming. It may not happen in our lifetimes, but we would all do well to prepare for it.

CHAPTER 17

The Whistleblower Report

If we become a united people, there is no doubt but we can with-stand the storms which threaten us. United we stand. United we are formidable.[1]

—Abigail Adams

During the summer of 2022, authorities began relaxing various restrictions related to the COVID-19 pandemic. Mask mandates were lifted in many places across the country. In a *60 Minutes* interview aired on September 18, 2022, President Biden declared that "the pandemic is over."[2] However, Secretary of Health and Human Affairs Xavier Becerra, continued to extend the public health emergency for COVID-19, doing so twice after President Biden's declaration with the latest extension on January 11, 2023.[3] The COVID-19 public health emergency, originally declared in January 2020, has been "extended" in ninety day increments twelve times. Many Americans may be confused by this extension in light of the government's own actions indicating that the emergency was over. What does the declaration of a public health emergency gain the United States, and why would officials continue extending that emergency?

The law requires the declaration of a public health emergency for our government to authorize any emergency use products. The "emergency" is what enables "emergency use" authorization of products that have not been licensed. All biological products must go through an approval process managed by the FDA. Without a declared public emergency, biological products have much more stringent approval requirements. As we continued seeing authorities push EUA products on unwilling recipients and then punish refusers, we realized that the entire basis for the EUA vaccines originated with the declared public health emergency. We tracked the emergency extensions as closely as we tracked the unavailability of fully FDA-licensed COVID-19 vaccines. Whenever the secretary of health and human services extends a declared public health emergency, the extension is good for ninety days. As each deadline approached, many of us prayed that the HHS Secretary would not extend the emergency so that the EUA product nightmare could finally be over. Each extension dashed our hopes and gave EUA products continued legal standing. This legal standing allowed medical providers to administer these EUA products, legally to the willing and illegally to the coerced and those not informed of the risks.

As the never-ending emergency dragged on, I felt like I was in a reconstruction of an eerily familiar science fiction story. Legislators essentially voted to grant the Galactic Supreme Chancellor limitless emergency powers to deal with an ongoing crisis. Never mind that the "crisis" had been the creation of the Supreme Chancellor himself. An unelected political appointee, the HHS Secretary was extending an "emergency"—a crisis our own government likely helped create—to maintain emergency powers.

How long could this extension of powers go on? The president of the United States had declared the pandemic over. What purpose were the emergency powers now serving? I could not help wondering when those governing us would attempt to "dissolve the Senate" or rewrite the Constitution to get rid of those "pesky inalienable rights." The sad reality is that under our current laws the administration has the ability to declare a never-ending public health emergency and then trample the rights of American citizens in the name of public health.

The public health emergency declaration places nearly all risks related to the receipt of an emergency use product directly on the recipient. Congress provides complete liability protection against any claim of loss for all persons and entities who are involved in the manufacture, distribution, planning, or administration of those products. The applicable law under Title 42 USC § 247d-6d(a)2(A) defines loss very broadly, listing everything from death to property loss from business interruptions. Even emotional injury caused by an EUA product is covered by complete liability protection. Persons and entities covered by liability protections include product developers, manufacturers, and vaccine administrators, as well as all related governmental personnel at the local, state, and federal levels, including members of Congress and the DoD.

Accepting administration of an emergency use product means the individual accepts nearly all the health, legal, financial, and medical risks arising from that product. (The Countermeasure Injury Compensation Program [CICP], a government program, is supposed to provide financial compensation for the injured, but we shall shortly see that it does not). This is the most important reason that recipients of EUA products must have the right to refuse. If one had no right to refuse an experimental product when the developers and manufacturers of that product are not incentivized to produce a safe product, an individual has become the equivalent of a lab rat, bred and sometimes murdered for the inhumane and deviant curiosity of "experts." The Nuremberg Code was developed to ensure such criminal activity could never happen again. Yet, our government and medical establishment have been party to a new dawn of criminally inhumane violations of the Nuremberg Code.

Injured recipients (or their families, in the event of death) who voluntarily receive an EUA product only have one thoroughly inadequate legal avenue to recoup their losses: filing a compensation claim through the CICP. As of December 1, 2022, the CICP had received 10,646 COVID-19 vaccine-related claims.[4] Of those 10,646 claimants, HHS had only approved payment to one individual, and that was not even a full payment. The Health Resources and Services Administration admits that this single claimant is still awaiting the rest of their payment due to "additional expenses to be calculated for this claim."[5]

Due to complete liability protections during declared emergencies, neither the executive branch of government, nor any manufacturer, developer, producer, or administrator of covered products has any incentive to ensure the safety or efficacy of the products they are providing. Congress must act to correct this loophole in the law. The legislature must put reasonable controls and limits on the power to declare and extend a declared public health emergency, or the medical freedom crisis will never end. If Congress does not act, I think it highly likely that new health threats will "emerge" the moment the previous one subsides. The federal government will then continue exercising endless emergency powers, and subsequent encroachments on individual rights will continue unabated.

Months before President Biden's public acknowledgement that the pandemic was over, the DOD's unlawful misrepresentation of interchangeability had already begun to fail in federal court. The terms *authorized*, *approved*, and *licensed* refer to different stages in the biologic licensing process dictated by federal law. Violations of that process are criminal, and the executive branch is the entity responsible for ensuring compliance with those laws.

In order to continue propping up the EUA products as permissible substitutes for the mandated FDA-licensed vaccine, the DoD and DOJ came up with a new rationalization to foist upon federal judges. They alleged that the Pfizer EUA vaccines were compliant with Biologics License Application (BLA) requirements. They coined the term "BLA-Compliant" in an apparent effort to argue that mandating an EUA product was lawful by conflating a product meeting certain license application requirements with actually receiving a license. Unlike the terms *authorized*, *approved*, and *licensed*, the term *BLA-Compliant* does not have any legal standing within the laws governing biological products.

Among the BLA requirements is an obligation to properly label biologic products. Not only did the EUA COVID-19 vaccines not receive a license, but these products were not even compliant with BLA requirements due to the labels not matching the BLA-approved product label. The BLA approved product label is and must be the FDA-licensed product label or it is not "BLA-Compliant." If the label says "EUA" it is not BLA-Compliant.

The proper labeling of an FDA-licensed product cannot be waived. Proper labeling for biologic products is so important that there are specific laws that govern labeling of FDA-approved products. 42 USC § 262(b) states that "No person shall falsely label or mark any package or container of any biological product or alter any label or mark on the package or container of the biological product so as to falsify the label or mark." The penalties for such violations are stated in 42 USC § 262(f): "Any person who shall violate, or aid or abet in violating, any of the provisions of this section shall be punished upon conviction by a fine not exceeding $500 or by imprisonment not exceeding one year, or by both such fine and imprisonment."

It is also important to note that fraud, including any fraud in labeling, voids liability protections and consent agreements. Arguing in court that an EUA-labeled product is legally equivalent to a fully FDA-licensed product is false, but it's not the same as the crime of false labeling. By this point in time, we had not seen any attempts at fraudulent labeling of EUA products, but we were concerned that it was a possibility and kept an eye out for it.

The "BLA-Compliant" effort by the DOJ and DoD quickly petered out due to the critical fact that the EUA products in question were *not* compliant with the laws governing labeling. An EUA product, labeled as an EUA product, cannot be mandated as if it were the FDA-licensed product. The labeling is not compliant with the law. This led to the next DOJ and

DoD attempt to rationalize the mandate of an EUA product to unwilling recipients, which we dubbed the "Comirnaty-labeled" effort. We began to see, in this effort, possible indications that the laws governing FDA-licensed product labeling were not being followed. The "Comirnaty-labeled" effort was initiated in May 2022 as part of the *Coker v. Austin* federal court case.

The *Coker* case was based, in part, on the lack of available FDA-licensed vaccines. Thousands of military members, including myself, confirmed the lack of FDA-licensed COVID-19 vaccines at military treatment facilities, pharmacies, and clinics across the country. An April 2022 Freedom of Information Act response from the Defense Health Agency (DHA) also confirmed that the DHA had no record of Comirnaty COVID-19 vaccines being ordered, received, in stock, available, or administered to *any* service member by *any* service branch (Army, Navy, Marine Corps, Air Force, or Coast Guard).[6]

The very next month, the DOJ, in an attempt to undermine the *Coker* plaintiffs standing in court, claimed that a fully FDA-licensed product was being made available. In a May 20, 2022 *Coker v. Austin* filing, the DOJ asserted that "While they may believe that FDA-approved vaccines are 'not available,' the Comirnaty-labeled vaccine is in fact available for DoD to order as of today's date."[7] Shortly thereafter, "Comirnaty-labeled" products began appearing in very limited quantities on certain military installations.

The sudden appearance of the alleged fully licensed product should have triggered another event we had been waiting for. In accordance with the law 21 U.S.C. § 360bbb-3(c), the HHS Secretary may only authorize a product for emergency use if a fully FDA-licensed product is not available. The HHS Secretary is further obligated by 21 USC § 360bbb-3(g) to review the progress made by fully licensed products and potentially revoke a product's emergency authorization if a fully licensed product becomes available. If the "Comirnaty-labeled" vaccine were truly the FDA-licensed Comirnaty, the HHS Secretary should have revoked EUA authorizations for all COVID-19 EUA vaccines. The never-ending emergency declarations, plus the still un-revoked Emergency Use Authorizations enabled the unapproved EUA biological products to remain on the market.

The HHS Secretary had ample opportunities to end the emergency. He instead continued to declare that the United States was in its longest public health emergency in history. When the first FDA-licensed COVID-19 vaccine was approved on August 23, 2021, the EUA authorizations should have been revoked. The HHS Secretary did not revoke them then. When the alleged FDA-licensed products were finally made available after

May 20, 2022, the HHS secretary still did not revoke the Emergency Use Authorizations. We have not been able to find any reasonable, coherent, or ethical justification for any of these actions by the HHS secretary. The only justification we could think of was the highly unethical extension of the emergency for the sole purpose of extending liability protections until the COVID-19 vaccines could be added to the CDC's recommended childhood vaccine schedule, at which point the current laws make liability protections permanent for vaccine manufacturers regardless of any "emergency."

We also found it very odd that the DOJ and the DoD only referred to this newly available product as "Comirnaty-labeled," rather than simply calling it Comirnaty. As Lieutenant Colonel Jon Cheek is fond of saying, "we don't call Coca-Cola, 'Coke-labeled' soda." We were deeply concerned that this terminology meant that our government had simply slapped a Comirnaty label on a product that did not follow all of the BLA requirements that would be followed by an actual licensed product. If this were taking place it would be fraud, and a violation of 42 USC § 262(b).

First Lieutenant Mark Bashaw, still attempting to do his job as a public health officer, had obtained access to the CDC's COVID-19 Vaccine Lot Number and Expiration Date Report Database. This database had originally noted that "These files contain all lots for COVID-19 vaccines made available under Emergency Use Authorization (EUA) for distribution in the United States." He observed that one of the new "Comirnaty-labeled" products, FW1331, was also showing up in the database list of COVID-19 vaccines "made available under Emergency Use Authorization." Bashaw detailed this in a sworn affidavit supporting Senator Ron Johnson's investigation into the safety and effectiveness of the COVID-19 vaccines.[8] He also detailed a number of other concerning pieces of information for Senator Johnson, including the abnormally high number of deaths, injuries, and adverse reactions to the COVID-19 vaccines within the Vaccine Adverse Event Reporting System (VAERS).

The CDC did not address the alarming safety signal identified in VAER but did, ironically, update the description of their Lot Number and Expiration Date Report Database to say, "These files contain lot numbers for COVID-19 vaccines made available under *either FDA Biologics License Application (BLA) or FDA Emergency Use Authorization (EUA)*."[9]

Knowing that the government had lied about so many significant purported facts related to COVID-19 vaccines, we were also curious and concerned about the new "Comirnaty-labeled" product's location of manufacture. The original August 23, 2021, BLA Approval letter for Comirnaty specified the manufacture location as either Puurs, Belgium, or Kalamazoo, Michigan.[10] The December 16, 2021 approval letter licensing the new 30-microgram dose formulation of Comirnaty removed Kalamazoo, Michigan, and specified that only the Pfizer manufacturing facility in Puurs, Belgium, would manufacture Comirnaty. Per the sworn testimony provided by whistleblower Lieutenant Chad Coppin, Pfizer admitted that Comirnaty Lot Number FW1331 was actually manufactured in France, not in the approved facility in Belgium.[11] The product apparently went through Kalamazoo, Michigan, on its way to Ft. Dietrick and into the Defense Health Agency system. However, based on the evidence provided, one of only three possibilities must be true: The reported manufacturing location was incorrect, the location was not in compliance with FDA licensing requirements, or the labeling was falsified at some point in the supply chain.

By August 2022, we had come a long way since the licensed COVID-19 vaccine had first been mandated. We had made it through the pandemic of fear, and although more than 8,400 service members had been kicked out over their resistance, many of us had been able to survive the subsequent coercion and abuse. When service members presented evidence of injuries and impermissible levels of risk from the COVID-19 vaccines, they were treated with the "Nuremberg Shrug" and endless repetitions of the "safe and effective" refrain. When service members could not find the licensed COVID-19 vaccine, senior military leaders created and promulgated a false "interchangeability" narrative and fraudulently used that narrative to deceive service members into thinking they had no right to refuse EUA products. They quickly moved to the false BLA-compliant ploy before finally producing a "Comirnaty-labeled" product.

All the while, the services were conducting unlawful religious discrimination against those whose consciences did not permit them to get a COVID-19 vaccine. They unlawfully denied religious accommodation requests and derailed the careers of those who fulfilled their duty to disobey unlawful orders. Service members and their families were harmed, injured, and abused. Appeals to commanders for help were ignored. Appeals to Congress for investigations revealed only a few willing to stand up and fight for service members and their rights. Recruiting and retention cratered to historic lows, with significant impacts to readiness. Mainstream narratives and congressional testimonies evaded the sad reality that the COVID-19 vaccine mandate was the centerpiece of an ideology that resulted in the destruction of American faith and confidence in what had previously been the most trusted American institution just four years ago.[12]

At the time, I was already working on the complaint against the naval inspector general, Admiral Fuller for his part in covering up unlawfulness by Admiral Grady and Admiral Nowell. I was beyond frustrated with everything I had seen and experienced in the last year. Most of our internal complaints and our reports to Congress were typically focused on one area, but the scope of DoD unlawfulness had grown massive. The DOJ was now apparently grasping for some foothold for the military vaccine mandate that could make them appear to be on the right side of the law.

An idea dawned on me based on First Lieutenant Bashaw's and Lieutenant Coppin's reports to Senator Ron Johnson that we should document the most critical aspects of the constitutional, legal, and readiness violations being committed by the DoD, write the story up in narrative form, and submit it to every member of the House and Senate. I knew I would need some help, so I pitched the idea to Mark Bashaw with a proposal that I would write it up, and he or I could be the individual to sign and submit it to Congress. Mark knew it needed to be bigger than that and proposed expanding the signatories beyond us. We settled on getting an officer or two from every military department plus Lieutenant Coppin as the Coast Guard representative.

I began asking around for team members who wanted to contribute to our newest project, but knew I had to start with Air Force Colonel JJ McAfee. In violation of multiple military regulations, JJ's chain of command almost immediately reassigned him following the submission of his own religious accommodation request. Rather than accept reassignment, he requested to retire, but was denied twice. His third retirement request was completely ignored. Despite personally enduring isolation and acts of reprisal at the hands of his chain of command, JJ encouraged those harboring anger to rise above the abuse and always act in love. JJ was an incredible leader to so many of us throughout the mandate, and I knew we needed his influence with this latest project. He was quick to say yes when I asked him to be the headlining signature on what we were beginning to call The Whistleblower Report. In addition to diversity of services, we also wanted to get some diversity of talent, including researchers, writers, legal experts, and of course Mark Bashaw and Chad Coppin as whistleblowers. We ended up with five additional officers: Army Lieutenant Colonel Jon Cheek, Navy Commander Liv Degenkolb, Air Force Major David Beckerman, Navy JAG Lieutenant Commander Patrick Wier, and Marine Corps Captain Joshua Hoppe.

There was something especially meaningful to me about having nine members of the whistleblower team. I had been reading a number of J.R.R.

Tolkien books at the time and I began thinking of our team as a contemporary version of the fellowship of the ring. Tolkien's fellowship of nine companions had been called to end a great conflict by bearing the ring of power into the heart of enemy territory and destroying it in the fire where it had been created. Our "Fellowship of the Whistle," as I began calling our team, was called to carry critical information into the heart of enemy territory and expose it to the fire of public scrutiny. Our initial brainstorming session resulted in a two-pronged effort. First, we would write the Whistleblower Report to Congress with only the most critical evidence and hard-hitting facts about the unlawfulness occurring. Then we would write and circulate a one-page petition with concise highlights from the Whistleblower Report and a link to our report and evidence.

We finished and signed our Whistleblower Report on August 15, 2022. We submitted it that day to congressional office email addresses and asked our extended network to get it to all of their representatives in Congress. Members of our network also forwarded the Whistleblower Report to news contacts resulting in stories across social media. Senator Ron Johnson, who had ongoing investigations on multiple related fronts, acted quickly on our latest information. With an August 18, 2022, letter to the DoD, FDA, and CDC, Senator Johnson asked very pointed questions about vaccine lot number FW1331 and the COVID-19 vaccines in general.[13] He included our Whistleblower Report as Enclosure 1 of his letter. Mainstream news outlets began picking up the story over the next several weeks, with a number of articles citing our Whistleblower Report.

The second blow of our one-two punch was the online petition. After drafting the one-page petition, we selected an online platform and posted both the one-page petition and the August 15 Whistleblower Report for further dissemination. Our petition was a call to Congress, on behalf of all Americans, to promptly investigate the illegality surrounding the military vaccine mandate and to hold accountable those found to have acted unlawfully. Captain Josh Hoppe took command of the petition campaign and began posting regular public updates. He also began engaging on social media to spread the truth about what was happening in the military and began submitting regular updates to Congress as the number of signatories to the petition grew.

While the petition helped with congressional awareness, the human rights organization Truth for Health Foundation helped significantly in getting our message out to the public. Truth for Health was already a powerful ally, supporting and helping defend service members fighting for their rights and for the Constitution. In addition to providing legal defense grants, the foundation used their radio and online platforms to call attention to our fight and to the harms perpetrated against so many service members. When our Fellowship submitted the Whistleblower Report on August 15, 2022, Truth for Health posted it on their website and amplified our message.

Truth for Health's founder, president, and CEO, Elizabeth Lee Vliet, MD, also had the idea of creating a new daily radio program to broadcast the messages of whistleblowers across the country. Dr. Vliet, supported by the foundations donors, created a series of Whistleblower Reports from numerous industries including medicine, military, law, ministry/faith, pharma/vaccines, environment, and education. She then purchased airtime on America Out Load Radio to create a daily "WeThePeople voice" for whistleblowers to expose injustice, deception, and harm, but also to demonstrate the hope and explore solutions for restoring our core constitutional principles. A growing group of military volunteers, including Dr. Pete Chambers, Liv Degenkolb, Brad Lee, Dixon Brown, Mike Gary, David Beckerman, and Brandi King now appear on and help host her new show. Our Whistleblower Report was also the inspiration for the title of the new Truth for Health radio program, which Dr. Vliet trademarked as the *Whistleblower Report*^{TM}.[14]

For many of us, the failure of DoD leadership and Congress to provide meaningful redress meant we had exhausted all non-judicial means of defending the Constitution against internal lawlessness. The favorable reaction to our August 15 Whistleblower Report was all the confirmation we needed that the only audience that mattered now was the American people; we could harness the power of free speech and gather the necessary public outcry required to effect change. Many DoD leaders had dug their heels in against the rule of law, military readiness, and basic common sense. They were not going to listen to us. Congress would only move in small increments and at the whim of politics. We would continue working with and informing Congress, but most members of Congress needed to see that the issues mattered to the majority of their constituents to be motivated to do anything. Therefore, the only audience that truly mattered anymore was the American people.

Personally, I needed Mark Bashaw to remind me of the wisdom in the maxim that our Founding Fathers relied on: *United we stand; divided we fall.* I had been a lone service member fighting back against DoD unlawfulness with my written efforts. My first whistleblower memorandum about religious discrimination to the House and Senate Armed Services Committees on January 7, 2022, was signed by me alone. I had regrettably not thought to join forces with others when submitting it. The HASC and SASC mostly ignored it. While some media entities were interested in reporting on my being fired, they did not show the same interest in my January 7, 2022, Congressional memorandum regarding religious discrimination.

Fortunately, I continue to be surrounded by incredibly insightful and powerful defenders of the Constitution, whose advice and inputs have made all our efforts more effective. When our fellowship stood together, and brought our concerns to bear with one voice, the resulting impact was much greater than any of us could achieve alone, greater even than I had ever anticipated. As Abigail Adams once wrote, "United we stand. United we are formidable."

CHAPTER 18

Awakening the Sleeping Giant of Liberty

It is now universally acknowledged that we are, and must be independent states. But still objections are made to a Declaration of it. It is said, that such a Declaration, will arouse and unite Great Britain. But are they not already aroused and united, as much as they will be? Will not such a Declaration, arouse and unite the Friends of Liberty, the few who are left, in opposition to the present system?[1]
—John Adams, June 23, 1776

Throughout 2021 and 2022, as we submitted our various internal memorandums, reports, and requests, we tried to find "friends of liberty" in the senior ranks of the US military and found very few. Our leadership mostly ignored our communications. When they chose to respond, it was often with retribution, resulting in various harms. Through the courage of hundreds of military plaintiffs and the talents of some incredible legal teams, we began securing some early legal wins after taking the fight to federal court. These wins, although successful in protecting service members from separation, were largely unsuccessful in awakening the American public to the unlawfulness and harms being perpetrated within the military. A number of factors, including significant censorship from major tech firms, kept the American people largely in the dark regarding the actions of domestic threats to the liberty and to our Constitution. We knew we had to do more in the public sphere.

We then saturated Congress with our petitions and requests for investigations. We found only a courageous few in Congress willing to take action on our information. These courageous few had very little success in reigning in the DoD's unlawfulness and were constantly being dragged into partisan political fights. This was very unfortunate from my perspective. I wanted nothing to do with partisan politics. I just wanted our leadership to follow the law. The vast majority of us couldn't care less which political party was in power, provided that those in power followed both the law of the land and the law of God. Our efforts with Congress did not lead to much success until we took the more grassroots and indirect approach we did with our Whistleblower Report.

Although there were a few journalists who had been fighting alongside us since the beginning, the vast majority of media personalities would not engage on the censorship or medical freedom issues, and generally shied away from challenging the prevailing narrative. Our Whistleblower Report and petition seemed to finally crack the code by attempting to gather a groundswell of support from "the people" in addition to appealing to Congress. It seemed that the Whistleblower Report itself was also more compelling to journalists than our previous efforts had been, not because of what we wrote or that we had finally connected all the major dots in a single document.

We had been saying parts of the same message for nearly a year, and mainstream media was not particularly interested. I believe there was so much more interest in the Whistleblower Report because we signed it as a group of officers from every military department and the Coast Guard. We signed it under our own names, not with pseudonyms or hidden identities. Like John Hancock when signing his name so visibly, we knew our adversaries would come for us, wanting to put a stop to the hope we inspired in others. I think it was this inspiration and hope that made our public stance so much more interesting to those who had mostly ignored our previous efforts.

Engaging directly with the American people became one of our most important missions and our final vector of action. We knew we had to awaken the American people to what was happening in the US military while, at the same time, hopefully rekindling in them an unquenchable yearning for liberty. The Whistleblower Report became a springboard to this most important of missions. Many of our nine signers were invited to do interviews for news articles, radio shows, and podcasts.

With so many modern distractions and readily available pleasures in our world, we would need help breaking through to the masses unaware of the danger our country was in. There had been journalists and media personalities fighting tirelessly for free speech since the beginning of the pandemic. They courageously took on difficult stories, especially stories that went against the prevailing narrative. Many of them were also willing to write about the military vaccine mandate and the critical readiness situation our leadership had put us in.

Our efforts to reach the American people would not have been impactful without the journalistic integrity and courage of individuals such as Kelly Laco, L. Todd Wood, John-Henry Westen, Bret Weinstein, Kristina Wong, and many others. One freelance journalist, J. M. Phelps, published more than fifty articles in 2022 about the military vaccine mandate and the harms being done to service members and their families. Even Tucker Carlson lent his platform to our fight, giving service members a chance to tell millions of people why we were taking this stand. Navy Commander Jay Furman, Marine Corps Lieutenant Colonel Scott Duncan, Air Force Master Sergeant Nick Kupper, Attorney Davis Younts, and Coast Guard Rescue Swimmer Zachary Loesch all made appearances on Tucker Carlson.

Zach Loesch's story is interesting due to the very public nature of his success in an operational environment as an unvaccinated service member. When Hurricane Ian devastated Florida in September 2022, Zach and fellow unvaccinated Coast Guard rescue swimmers, Chad Watson, Ian Jobs, and Justin Bastow were sent with a team to support FEMA search and rescue efforts. The interagency operation successfully rescued over 3,500 people, thirty-four of which were personally handled by this group of unvaccinated rescue swimmers. Zach's efforts had such an impact that President Biden personally called him to express America's thanks for his heroism. Zach's story is positive proof that unvaccinated service members can and should be operating in all the operational environments they served in prior to the implementation of the military vaccine mandate.

The unvaccinated rescue swimmers' success in this challenging environment proved beyond a reasonable doubt that efforts to separate them and other service members like them did more harm than good. Efforts to needlessly discard such capable warriors when our country is suffering through one of the worst recruiting droughts in decades, can only be interpreted as a policy-based domestic threat to the Constitution and to military readiness. In the face of such overwhelming evidence, I can think of no other reasons for such self-destructive policies than pure political motivations and ambitious careerism.

My own efforts to reach the American people saw very limited success early on with only a few podcast and radio show appearances. But, in early August, as we were finalizing the Whistleblower Report, I received an invitation that would break down significant communication barriers for me. I was invited to be a speaker at the Gateway to Freedom Conference by Tom Stewart, a former Marine Corps pilot separated during the vaccine mandate purge. The conference, put on by Dr. Mollie James, was to take place in St. Louis later that month. With my family and I preparing for an upcoming PCS move, the Gateway to Freedom Conference team was kind enough to allow me to appear remotely via video. Having seen the agenda and speaker list, I knew this was an opportunity I could not pass up. The Gateway to Freedom Conference brought together dozens of the most powerful leaders in the Medical Freedom movement, including Dr. Peter McCullough, Dr. Robert Malone, and Mr. Ed Dowd.

The Gateway to Freedom Conference was also unique in that it was the first major medical freedom conference to incorporate a significant active-duty military presence. Dr. Mollie James understood the significant role the military was playing in the fight for medical freedom and dedicated almost an entire day of the conference to military speakers and military-specific issues. I was very gratified to see some familiar military names on the Gateway to Freedom speaker schedule including Lieutenant Colonel Scott Duncan, Captain Grant Smith, Major David Beckerman, Dr. Sam Sigoloff, and Dr. Pete Chambers. A United States Air Force Academy professor fired for refusing the COVID-19 vaccine, Retired Lieutenant Colonel Sandy Miarecki, PhD, was scheduled to speak later on the same day I was.

On the day of the conference, I had the great honor of following immediately after Dale Saran. In 2000, during the anthrax vaccine debacle, Dale Saran was one of the JAG lawyers defending anthrax vaccine refusers at separation boards and court-martials. When the COVID-19 vaccine was mandated in the military, Dale, now a civilian lawyer, picked up where he left off and now has nearly a thousand military plaintiffs in cases against the federal government. Dale had fought alongside Colonel Tom Rempfer (now retired) during the anthrax vaccine fight. Both Tom and Dale were instrumental in the fight that resulted in the permanent injunction issued by a 2004 US District Court prohibiting the government from administering investigational drugs to service members without their consent.[2] The current EUA laws are a legacy from the efforts of Dale Saran, Tom Rempfer, and so many others who pushed back against earlier unlawful attempts by the Department of Defense to force experimental biological products on military personnel.

I kept my remarks very brief during my talk at the Gateway to Freedom Conference. I only wanted to make three points. First, I talked about the importance of networking and the connections we made with others fighting for the Constitution. As an example, I highlighted the relationship I built with the Navy SEAL master chief whose wisdom had been so instrumental to my own efforts. I also told elements of the inside story about the Navy's class-wide preliminary injunction and highlighted those individuals whose efforts made it a reality. Lastly, and most importantly, I shared my thoughts on how we can ultimately win this fight. I think victory will not come with a simple reversal of the mandates and a full repairing of the harms caused. We have a deeper problem in our country. We will have to continue fighting tyranny and government overreach in all areas of our lives until we take our country back at the local, state, and federal level. The two-party system is broken and enables entrenched operators to continue driving us further from the intent of our Founding Fathers.

I shared with the conference attendees an effort championed by the newly formed America's Veteran Party, a nonprofit founded by Dr. Grant Smith focused on supporting veterans and supporting the Constitution by getting constitutionalist military veterans elected to political office. Veterans who put their lives and livelihoods on the line to fight for their basic constitutional rights are the only people I would trust to not become corrupted by the money, power, and benefits that federal-level politicians expect as a matter of course. It is also these constitutionalist veterans I would trust to put congressional term limits in place and dismantle the unnecessary benefits packages that even former one-term members of Congress receive.

America's Veteran Party has already endorsed several candidates that I know, either from serving with them or through our extended medical freedom network. Mara Macie, a military spouse and mother of four who has several years of homeschooling experience, ran for federal office in Florida's 5th congressional district on an America-First platform. She championed medical freedom and wanted to vote in term limits for members of Congress. Jordan Karr, an Air Force veteran separated over the vaccine mandate, ran for a Florida state office in 2022 as well. Both lost their bids, but they are the exact type of candidates I believe we need in office, and I hope they continue running.

I actually served with Chief Petty Officer Morgan Martinez, the third candidate endorsed by America's Veteran Party that I know personally.

Morgan, now a commissioned officer in the Navy Reserve, won his race and was elected county commissioner of Washakie County in Wyoming. As I explained to the Gateway to Freedom Conference attendees, we will not win our country back until we have constitutionalist candidates in a majority of public offices at every level of government. It began with a single courageous decision, but those who stand for medical freedom and for the Constitution are the perfect candidates to begin this movement.

The domestic enemies of the Constitution are already motivated and united. Taking a public stand against them is absolutely required, and not just because we took an oath. They want to isolate, censor, and silence all dissent by any means necessary, including means that clearly violate the First Amendment rights of Americans. A public testimony is required to give hope to others as well as to motivate and unite those who would oppose the dismantling of the Constitution occurring all around us.

In the months leading up to the signing of the Declaration of Independence, many objected to making such a public declaration out of fear that it would provoke Great Britain to further wrath against the colonies. John Adams understood, however, that taking no action would be a much greater risk than a public declaration of the colonies' intent to separate from the British crown. The British were *already* provoked and trampling the rights of our colonial forbears as a result. The Declaration of Independence served not just to inform the British, but also to unite the colonies in the hope and promise of liberty.

We face a nearly identical challenge today. The enemies of liberty are tyrannizing the American people and trampling their God-given rights. They are motivated to see their plans to whatever end they have in store by any means necessary. The inexcusable encroachments on individual medical freedoms and other basic rights must be opposed, because history tells us that if they aren't it will end badly for all. We don't know exactly how bad it could get, but we can be sure that the belief systems that motivate such will-ful encroachments on our rights will only be emboldened by compliance.

Because the forces arrayed against us are so coordinated and motivated, we also cannot afford to fight as isolated individuals unconnected to the wider public. We must publicly unite and challenge those who would attempt to trample our rights and dismantle our Constitution. There are many "friends of liberty" that must be awakened in order to address this existential threat to our Republic. If we remain silent and do not speak out publicly, we may not be able to find those who need that small nudge of courage to act. Therefore, following the example set by our Founding Fathers, we declare our opposition to those tyrants and oppressors who would put compliance with unconstitutional edicts over our God-given rights guaranteed by the Constitution. This declaration will give those watching from the outer edges and those on the fence about whether to join us courage and hope.

When leaving the Constitutional Convention over two hundred years ago, Benjamin Franklin is reported to have fielded a question about what form of government the convention established. Franklin responded by stating, "A republic, if you can keep it." If we are to be successful in keeping our constitutional republic, we must unite in opposition to the enemies of liberty, relentlessly focused on tearing down individual liberties and constitutional rights. We call on all Americans and all people of good will to join us in defending the Constitution against all enemies foreign and domestic.

CHAPTER 19

The Call to Arms

The troops are constantly to have one day's provisions on hand, ready cooked—The officers are to pay particular attention to this, and consider it as a standing rule, that if they are suddenly called to arms, the men may not be distressed.[1]
—General George Washington, General Orders,
December 4, 1777

This is the part of the book where the author appeals to the love of freedom, sometimes buried dormant, in the heart of every man, woman, and child. This appeal could be in the form of an impassioned call to muster the forces of freedom for a great offensive against the decaying effect of tyranny on a people long oppressed. However, I choose to make a different type of call to arms. It is true that the fight against tyranny must be uncompromising and must be very public. Yet, this fight should not look like the fifth-generation warfare campaigns of our enemies, overtly focused on social engineering, misinformation, and deception. Instead, our fight must be a fight to remain faithful. The call to arms I have to offer is actually a call to faithfulness.

So many leaders, acting in good faith, rationalized their way into compliance. Many went against their better judgement in accepting a COVID-19 vaccine or in demanding that others do so. Our call to faithfulness is first and foremost a call to remain faithful to conscience. If something doesn't feel right, it probably isn't. Don't press forward on the word of others. Make sure you understand every aspect of what you are being asked to do before doing it. Make sure you understand what you are asking of others before you ask it. The current fight over the competing interests of individual medical freedom and a government program intent on 100 percent compliance is likely just the beginning. The next fight may be over something much more sinister. What if you are asked to round up your unvaccinated and non-compliant neighbors and friends? What if you are asked to put them in train cars to designated quarantine zones? Would this be a lawful order? Would you be defending the Constitution by complying with these orders? I believe resisting such orders would be the only way to keep my own conscience clear.

In addition to being faithful to conscience, we must also be faithful to the Constitution. The Constitution is more than just a meticulously preserved piece of colonial-era parchment. The Constitution was meant to both establish and, more importantly, limit a newly formed government. The limiting of that government is an often overlooked but critical element of what makes our Constitution so important to the defense of individual rights. Defending the Constitution also does not mean defending institutions that were established by the government when those institutions are at odds with the Constitution. Defending the Constitution is most critically a defense of individual rights against those who would encroach on those rights or attempt to take those rights away.

We military members are trained to defend against a foreign power that may attempt to trample, limit, or take away our rights. This is, however, just one aspect of what we have sworn an oath to defend against. The oath we took is to defend the Constitution against *all* enemies, foreign and domestic. Domestic threats include more than just those who want to overthrow our Constitution and establish some other way of life. The greatest domestic threat we currently face is from individuals who have engaged the complicity and compliance of those in power to encroach on individual medical freedom rights guaranteed by the Constitution.

In his December 4, 1777 General Orders, General George Washington required his soldiers to have at least one day of provisions on hand so they would be ready if given the call to arms. We must also be ready when called on to fight for individual constitutional rights. The "provision" each of us must have on hand is personal courage. Courage is a virtue and must be regularly exercised in order to strengthen it. The lack of practice in a particular virtue leads to the habitual use of its opposite vice. If you never practice courage, for example, you will develop a habit of cowardice. Similarly, if you never practice moral courage in the small things, it is likely you will not be able to make the difficult moral decision when the stakes are much higher.

For military members a simple exercising of courage in the small things could be telling a superior the true limitations of a mission or the true impact a new requirement would have on already-overworked troops. So often commanders become "yes-men," who find a way to "get the job done," rather than pushing back when things don't make sense. These are the people we have promoted for decades. As an Air Force C-130 pilot recently shared with me, "If you keep saying 'yes' when you shouldn't, you won't be able to say 'no' when you should." I found his quote to be a perfect example of what the lack of practice in the virtue of courage can do to a military leader.

We need to start prioritizing the promotion of independent thinkers and those who are willing to take morally courageous positions, especially when those stances are unpopular. Popularity will not win wars and saying only what a superior officer wants to hear, may help career progression, but it will also breed strategic complacency. Wars are lost by countries whose senior military ranks are full of automatons and "yes-men." The greatest military leaders in our nations' history thought outside the box to develop unique solutions to the warfighting challenges they faced. Many of them were also aggressively independent and tactically unpredictable; think Patton, Halsey, MacArthur, and Spruance. Independent critical thought at all levels of the chain of command is essential to achieving the warfighting advantages required in the unpredictable modern battlefield.

The final element of the call to faithfulness is to remain steadfast. We do not know where or how this battle is going to be fought in the future, but we do know it will continue to be fought. We must remain steadfast in our fight for individual constitutional rights. Failure to remain steadfast to the end is a failure in faithfulness. We must not allow threats of harm, or even the realization of those threats, to cast us from our destiny as heirs to the great gift of liberty.

We who held our ground against the military's illegal enforcement of the COVID-19 vaccine mandate took an oath to defend the Constitution and will remain faithful to that oath until our last breath. The domestic threats trampling individual constitutional rights have awakened and united a group of individuals who take that oath seriously. The sleeping giant of liberty has been provoked to fight back. We choose now to fight back, not with swords or any other armament, but with truth and faithfulness. We will not stand idly by while the Constitution is dismantled around us. Despite being behind enemy lines we have chosen to place faithfulness to our oaths, to our consciences, and to our country before all else. May future generations learn from these lessons as we have learned from those who founded our nation and handed down the Constitution that we swear to continue defending.

Epilogue

The Revolution was in the Minds of the People, and in the Union of the Colonies, both of which were accomplished, before Hostilities commenced. This Revolution and Union were gradually forming from the year 1760 to 1775. The Records of the British Government and the Records of all the thirteen Colonies, and the Pamphlets Newspapers and handbills of both Parties must be examined and the essence extracted before a correct history can be written of the American Revolution. [1]

—John Adams, 1815

On December 23, 2022, President Biden signed the 2023 National Defense Authorization Act (NDAA) into law. The NDAA included a provision ordering the DoD to repeal the COVID-19 vaccine mandate. The NDAA states that "Not later than 30 days after the date of the enactment of this Act, the Secretary of Defense shall rescind the mandate that members of the Armed Forces be vaccinated against COVID-19." While this may be a generally positive step in the right direction from a policy perspective, the congressionally directed end to the COVID-19 vaccine mandate did nothing to correct harms already done. Congress did not direct the DoD to reinstate the military members fired for refusing the COVID-19 vaccine, nor did they provide any of the benefits service members lost over the mandate.

This book is meant to record our actions for posterity's sake so that the American people may know and understand why so many of us took the stand we took. It is also important that we record what actions our government and our military leaders took in mandating and enforcing a COVID-19 vaccine mandate that has resulted in significant harms to individuals, families, and to our national security, with essentially zero measurable positive benefit, except for the profits of pharmaceutical companies. Nothing our government does in the future will change the unlawful actions that many leaders took during the year and a half that Secretary Lloyd Austin mandated the COVID-19 vaccine. The government can correct policy mistakes and leaders can work to rebuild trust, but they cannot rewrite history regardless of how hard they try. It is our duty to ensure that history does not forget the unlawful actions taken by so many military leaders and government officials.

As far as rebuilding trust, one of the best proposals I read came from Air War College assistant professor, Colonel Paul Vicars, PhD. In a December 16, 2022, article in *The American Conservative*, the fighter pilot and former training squadron commander offered several solutions to rebuild trust with service members who were still under threat of discharge before President Biden signed the 2023 NDAA. According to Colonel Vicars, the very first step to rebuilding trust would be for the *services* to "end the policy that is breaking the trust—the Covid vaccine mandate."[2]

Sadly, the services did not take this step on their own. The DoD's inaction and stubbornness forced Congress to take this action for them. The fact that outside institutions had to force the DoD to follow the law and end the mandate will only deepen the lack of trust that service members have for their leaders and the DoD. Service members were forced to go to the federal courts to get the DoD to stop violating basic First Amendment rights and federal law as established in the Religious Freedom Restoration Act. Then it was Congress, not the DoD, that had to fix the initial policy mistake and end the COVID-19 vaccine mandate. Service members are only going to see this as further proof that their leaders within the DoD cannot be trusted to do the right thing. The next time there is a conflict that needs resolution, service members are going to be even quicker to seek redress with the federal courts and Congress, now that the DoD has become so untrustworthy.

Colonel Vicars also offered several actions that military leaders can take to make things right with service members who had not yet been involuntarily discharged. The DoD, Vicars argues, should allow those service members who see the differences they have with the DoD as irreconcilable, to leave the service with full honors. For those who indicate a willingness to reconcile and continue service, Vicars argues that the harms they faced must be rectified as completely as possible. He writes that commanders and functional managers should be empowered to grant assignment preferences, prioritize training to recover lost training opportunities, and grant special assignments to get stalled careers back on track.[3]

These actions, however, require commanders who *want* to make things right with the service members they harmed. The vast majority of commanders are not proving to have any desire to repair harms. The way Air Force Captain Brennan Graves was treated immediately following the repeal of the military vaccine mandate is an important case study to understand the tone-deaf and career-focused approach senior leaders are taking with their victims.

Brennan Graves is an E-6B flight test engineer with a heart valve regurgitation that he has had since childhood. This condition, which runs in his family, causes a slight reduction of blood flow through one of the valves in his heart. Although this condition never impacted his ability to fly or perform his duties, the pharmaceutical companies' advertised COVID-19 vaccine risks caused Brennan enough concern to consult with a cardiologist. The cardiologist informed Brennan that a COVID-19 vaccine could be very dangerous for him, and he should not receive one under any circumstances.

Despite a letter from this cardiologist, military medical providers still denied his medical exemption request. When he filed a religious accommodation request, his squadron then discriminated against him and isolated him while his commander removed him from flight status. His commander then recommended disapproving his religious accommodation request while informing the commanding general that Captain Graves was a threat to the squadron.

Reading the writing on the wall, Brennan attempted to transfer to another unit where he could "get off the bench" and have an impact. His transfer requests to Test Pilot School, Squadron Officer School, and ROTC instructor duty were all denied multiple times. He then attempted to resign his commission and leave the Air Force, which was also denied. We had seen so many retirement and resignation request denials for unvaccinated service members over the past year that we developed a new phrase for it; malicious retention.

Captain Brennan Graves's malicious retention and the multiple denials of his transfer requests were only a small part of the harms perpetrated against him by his leadership. Brennan was a top performer before the COVID-19 vaccine mandate and had been a meritorious recipient of the Air Force's Global Strike Command (GSC) Company Grade Officer of the Year award. Yet when Brennan's Aerial Achievement Medal citation came up for review, his commander refused to approve it. Unlike vaccinated service members, Brennan was forced to submit to COVID-19 testing repeatedly. Over the course of his discrimination, isolation, and abuse, Brennan was tested for COVID-19 a total of 103 times, despite having natural immunity. Every single COVID-19 test that Brennan took came back negative. After his religious accommodation request came back denied, his commander attempted to award non-judicial punishment which Graves refused, demanding a court-martial instead. When Brennan was finally able to get a transfer, his commander denied his house hunting leave, a benefit that is almost universally approved for service members moving their families across the country.

After all these actions perpetrated against Brennan, the morning after the Congress-directed repeal of the vaccine mandate, Brennan's commander come right up to him and said, "Let's get you flying. The squadron and flight need you out there!" There was no attempt to repair the harms done to Brennan. Nor was there any apology for the horror that this commander had put Brennan and his family through. Tragically, Brennan's story is just one of hundreds of similar stories. It is obvious to me that senior military leadership will not hold themselves accountable nor are they willing to learn from their mistakes. Either Congress or the federal courts must hold these commanders accountable for their dereliction of duty and numerous violations of service members' constitutional rights.

The DoD did not end the mandate willingly and has certainly not moved to correct the many harms done to service members who exercised their rights to remain unvaccinated. To make matters worse, the DOJ almost immediately began scrambling to take advantage of the 2023 NDAA to the potential detriment of those harmed by the DoD. The DOJ began filing dismissal requests for the various lawsuits filed by military plaintiffs due to the repeal of the mandate. This appears to be a concerted and desperate effort to ensure that none of the lawsuits ever return a ruling that the DoD's actions were unlawful.

In some ways the Congressional action taken in the NDAA was so limited in scope that it may serve to make our fight even more difficult. In the 2023 NDAA, Congress did not acknowledge the unlawful actions taken by the DoD, nor did they attempt to force DoD leaders to repair the harm they caused to so many service members and their families. The DOJ is attempting to argue that with the mandate rescinded there is no longer a need for the lawsuits and they can be dismissed. In some ways, this would be the worst crime of all. If they are successful in these efforts, it is possible that neither definitive legal remedies to redress the harms done, nor clear court rulings on the lawfulness of the vaccine mandate itself will ever be made. If we are unable to obtain a clear precedent setting ruling, I fear the

DoD will be emboldened in their encroachment of service members' medical freedom rights. The next time the DoD engages in unlawful actions in the name of some future agenda, it will be all the more difficult to stand up to it and stop it.

The DoD and the federal government should make every effort to repair the irreparable harms inflicted on service members and their families. As established by federal court rulings, those who were coerced into violating their faith, consciences, and First Amendment rights are victims of irreparable harm. Others, smarter than I, will have to determine how our government is to repair the irreparable harms of those who were coerced into receiving a COVID-19 vaccine against their consciences and belief systems.

The same applies to those who endured some permanent physical damage or lost their lives due, even in part, to the COVID-19 vaccine mandate. These injuries and deaths are irreparable in the most tragic way. The survivors and the families of the lost are suffering. The federal government has a moral obligation to do everything it can to right these wrongs and ease that suffering. I operate in the hope that this moral obligation will be enough to prompt the required action. In case the federal government and certain military leaders continue failing to fulfill their moral obligations, we will continue pursuing federal court cases in order to turn their moral obligations to repair these harms into legal obligations.

John Adams noted in 1815 that their own revolution was accomplished in the "minds of the people" long before hostilities commenced. In our present situation, the roles of revolutionaries and defenders of the established system are reversed. The forces of overwhelming federal tyranny are the ones apparently fomenting an overthrow of the Constitution and rule of law, while the defenders of the American constitutional republic are the ones forced by their oaths and consciences to stand up to oppression and the trampling of basic human rights.

Future historians will write about this time and the actions we took during this battle. History is typically written by the winners, so I do not know how our actions will be seen and interpreted by future generations. I do know why I took the actions I did and what I intended by those actions. My intentions, shared by so many men and women of courage and conscience, were to fulfill my oath and defend the Constitution. This book represents my ongoing attempts to defend the Constitution and ensure that the winners of the current hostilities have a documented account of our resistance to lawlessness and tyranny.

John Adams noted that a correct history of the American Revolution could not occur without extracting the essence of the records of both the British Empire and the thirteen American colonies. So, it will be for us. No history of the current hostilities will be correct without records from both sides of this conflict. Whatever others write about us and our intentions, I pray that this account survives and becomes part of the history future generations study when seeking to learn lessons from our battle over medical freedom, constitutional rights, and the free exercise of religion in America.

Acknowledgments

A work like this is not possible without significant support, help, and guidance. I acknowledge first and foremost, that I am wholly inadequate to the magnitude of the implications and the wider importance of the conflict between federal overreach and basic individual constitutional, religious, and human rights. I spent a great deal of time in prayer for guidance throughout the writing of this book. This book is therefore dedicated first to Our Lady of Liberty, whose intercession before the throne of God I prayed for. Any failures of this book are mine, and all successes of this book and any good this book does in the world must be attributed to mercy, power, and majesty of Christ the King. I must also acknowledge my wife Mary Clare and our seven children. Thank you for being the constant tangible reminders of what this fight is all about for me. To my brothers Christopher and Jeffery, thank you for being my first reviewers. To my brother Matt, thank you for being the best chief petty officer I know. There is a reason sailors are taught to "ask the chief." Despite being my younger brother, I really look up to you and found your thoughts and assistance most edifying. To all the men and women of conscience I mention in this book, and those I was unable to mention, I thank you for your courage. All of us are comrades in arms and many of you have become very close personal friends. This book, and what I describe within it, would not be possible without your influence, your integrity, and your courage to act. To the American people, please know that if we can "keep our Constitutional Republic" as Benjamin Franklin desired, it will be due, in part, to the thousands of service members who sacrificed everything to choose the difficult right over the easy wrong these past two years.

Author's Note

The vast majority of my research into the Founding Fathers' writings, ideas, and concepts was done using Founders Online. Founders Online is a database of the writings of many of our Founding Fathers. Scholars gathered the Founding Fathers' papers and personal effects, meticulously researching, preserving, and digitizing them for future generations. They even note whose handwriting was used in situations where the author was dictating a letter to someone else, as General George Washington often did to his military aides, including Alexander Hamilton. To all the scholars who worked so tirelessly on the Founding Fathers archive project, I am incredibly grateful. Paper will decay, and hand-written ink will fade. It is my hope, however, that the digitized database of our Founding Fathers' writings can be preserved and shared in perpetuity.

Towards the end of the writing of this book, Elon Musk publicly released evidence of US Federal Government collusion with Twitter in the active censoring of free speech. It did not matter which political party was in power because the last two presidential administrations were apparently involved in this unlawful, unconstitutional, and tyrannical censorship effort. Most Americans knew this was happening, and the public revelation of the evidence should have come as no surprise. Many Americans were outraged that their 280-character ideas and concepts could be censored. My first thought, however, went to databases such as Founders Online. If our government, oligarchs, and technocrats can censor free speech with impunity, what is to then stop them from censoring, or removing our access to archives of the Founding Fathers' writings? We must continue to be vocal and stop any efforts to censor the speech and resources that give hope and inspiration to those fighting for the preservation of freedom.

Notes

Chapter 1

1. "From John Adams to Massachusetts Militia, 11 October 1798," Founders Online, National Archives, https://founders.archives.gov/documents/Adams/99-02-02-3102.
2. Office of the Secretary of Defense, "Leadership Stand-Down to Address Extremism in the Force," accessed November 8, 2022, https://media.defense.gov/2021/Feb/26/2002589872/-1/-1/1/LEADERSHIP-STAND-DOWN-FRAMEWORK.PDF.
3. Eleanor Watson, "Pentagon updates extremism policy and says nearly 100 service members engaged in extremist activity in 2021," CBS News, December 21, 2021, https://www.cbsnews.com/news/pentagon-extremism-military-policy-updates/.
4. U.S. Congress, Senate, Committee on Armed Services, Report 117–130, James M. Inhofe National Defense Authorization Act for Fiscal Year 2023: Report (to Accompany S. 4543), 117th Congress, 2d Session, July 18, 2022, 159–160, https://www.congress.gov/117/crpt/srpt130/CRPT-117srpt130.pdf.
5. History.com Editors, "My Lai Massacre," November 9, 2009, Updated October 4, 2022, https://www.history.com/topics/vietnam-war/my-lai-massacre-1.

Chapter 2

1. "To John Adams from Samuel Adams, 4 October 1790," Founders Online, National Archives, https://founders.archives.gov/documents/Adams/06-20-02-0247. [Original source: The Adams Papers, Papers of John Adams, vol. 20, June 1789–February 1791, ed. Sara Georgini, Sara Martin, R. M. Barlow, Gwen Fries, Amanda M. Norton, Neal E. Millikan, and Hobson Woodward (Cambridge, MA: Harvard University Press, 2020), pp. 417–419.
2. "Samuel Adams," The Belcher Foundation, accessed October 1, 2022, http://belcherfoundation.org/samuel_adams.htm. (This biography of Samuel Adams was originally adapted from William Allen, An American Biographical and Historical Dictionary, Containing an Account of the Lives, Characters, and Writings of the Most Eminent Persons in North America from Its First Discovery to the Present Time, and a Summary of the History of the Several Colonies and of the United States (Cambridge: William Hilliard, 1809), pp. 3–6.

3   Mark Puls, *Samuel Adams: Father of the American Revolution,* (New York: St. Martin's Press, 2007), Introduction, Kindle.

4   Eugene M. Del Papa, "The Royal Proclamation of 1763: Its Effect upon Virginia Land Companies," *The Virginia Magazine of History and Biography* 83, no. 4 (October 1975): 406–411.

5   "Samuel Adams's Instructions to Boston's Representatives," Samuel Adams Heritage Society, accessed October 2, 2022, http://www.samuel-adams-heritage.com/documents /samuel-adams-instructions-to-bostons-representatives.html.

6   Rebecca Beatrice Brooks, "What Caused the American Revolution," *History of Massachusetts Blog,* accessed October 1, 2022, https://historyofmassachusetts.org/what-caused-american-revolution/.

7   "The Great Debate," ConstitutionFacts.com, accessed October 2, 2022, https://www .constitutionfacts.com/us-articles-of-confederation/the-great-debate/.

Chapter 3

1   "Pennsylvania Assembly: Reply to the Governor, 11 November 1755," Founders Online, National Archives, https://founders.archives.gov/documents/Franklin/01-06-02-0107. [Original source: *The Papers of Benjamin Franklin,* vol. 6, April 1, 1755, through September 30, 1756, ed. Leonard W. Labaree (New Haven and London: Yale University Press, 1963), pp. 238–243.

2   Caroline Eisenhuth, "The Coercive (Intolerable) Acts of 1774," George Washington's Mount Vernon, accessed 20 Oct 2022, https://www.mountvernon.org/library /digitalhistory/digital-encyclopedia/article/the-coercive-intolerable-acts-of-1774/.

3   "Table of Laws Held Unconstitutional in Whole or in Part by the Supreme Court," Constitution Annotated: Analysis and Interpretation of the U.S. Constitution, accessed 4 Oct 2022, https://constitution.congress.gov/resources/unconstitutional-laws/.

4   "Table of Supreme Court Decision Overruled by Subsequent Decisions," Constitution Annotated: Analysis and Interpretation of the U.S. Constitution, accessed 18 Oct 2022, https://constitution.congress.gov/resources/decisions-overruled/.

5   James Colgrove and Ronald Bayer, "Manifold Restraints: Liberty, Public Health, and the Legacy of Jacobson v Massachusetts," *American Journal of Public Health* 95, no. 4 (April 2005): 571.

6   Josh Blackman, "The Irrepressible Myth of Jacobson v. Massachusetts," *Buffalo Law Review* 70, no. 1 (January 2022): 156–158.

7   Blackman, "The Irrepressible Myth of Jacobson v. Massachusetts," 152, 166–167.

8   Wendy K. Mariner, George J. Annas, and Leonard H. Glantz, "Jacobson v Massachusetts: It's Not Your Great-Great-Grandfather's Public Health Law," *American Journal of Public Health* 95, no. 4 (April 2005): 582.

9   Commonwealth v. Pear, 183 Mass. 242 (Mass. 1903), at 248.

10   James Colgrove and Ronald Bayer, "Manifold Restraints: Liberty, Public Health, and the Legacy of Jacobson v Massachusetts," *American Journal of Public Health* 95, no. 4 (April 2005): 571.

11   Jacobson v. Massachusetts, 197 U.S. 11 (1905), at 26.

12   Blackman, "The Irrepressible Myth of Jacobson v Massachusetts," 178.

13   Blackman, "The Irrepressible Myth of Jacobson v Massachusetts," 179.

14   Blackman, "The Irrepressible Myth of Jacobson v Massachusetts," 179–180.

15 Blackman, "The Irrepressible Myth of Jacobson v Massachusetts," 181–185.

16 Blackman, "The Irrepressible Myth of Jacobson v Massachusetts," 181.

17 Wendy K. Mariner, George J. Annas, and Leonard H. Glantz, "Jacobson v Massachusetts: It's Not Your Great-Great-Grandfather's Public Health Law," *American Journal of Public Health* 95, no. 4 (April 2005): 583.

18 Jacobson v. Massachusetts, 197 U.S. 11 (1905), at 39.

19 Adam Cohen, *Imbeciles: The Supreme Court, American Eugenics, and the Sterilization of Carrie Buck* (New York: Penguin Press, 2016), 317–318.

20 Buck v. Bell, 274 U.S. 200 (1927), at 207.

21 Cohen, *Imbeciles*, 2, 44–48.

22 Blackman, "The Irrepressible Myth of Jacobson v Massachusetts," 191.

23 Oliver Wendell Homes Jr. to Felix Frankfurter, Sept. 3, 1921, (6 years before the Buck v. Bell ruling) in *Holmes and Frankfurter: Their Correspondence, 1912-1934*, ed. Robert Mennell and Christine Compston (Hanover, NH University Press of New England, 1996), 125, as quoted in Adam Cohen, *Imbeciles: The Supreme Court, American Eugenics, and the Sterilization of Carrie Buck* (New York: Penguin Press, 2016), 242.

24 Buck v. Bell, 274 U.S. 200 (1927), at 207.

25 Buck v. Bell, 274 U.S. 200 (1927).

26 Blackman, "The Irrepressible Myth of Jacobson v Massachusetts," 193.

27 Cohen, *Imbeciles*, 19–21.

28 Cohen, *Imbeciles*, 21–22, 297.

29 Cohen, *Imbeciles*, 24–27.

30 Cohen, *Imbeciles*, 16.

31 Cohen, *Imbeciles*, 34–35.

32 Cohen, *Imbeciles*, 24, 292.

33 Cohen, *Imbeciles*, 319.

34 Blackman, "The Irrepressible Myth of Jacobson v Massachusetts," 135.

35 Blackman, "The Irrepressible Myth of Jacobson v Massachusetts," 269–270.

36 Cohen, *Imbeciles*, 2, 303.

Chapter 4

1 Dagobert D. Runes, ed., *The Selected Writings of Benjamin Rush* (New York: Philosophical Library, 1947), 315.

2 Congressional Research Service, Operation Warp Speed Contracts for COVID-19 Vaccines and Ancillary Vaccination Materials, by Simi V. Siddalingaiah. IN11560. March 1, 2021.

3 "What is Operation Warp Speed?" NIH: National Institute of Allergy and Infectious Diseases, June 1, 2020, https://www.niaid.nih.gov/grants-contracts/what-operation-warp-speed.

4 "Were the COVID-19 vaccines rushed? Here's how the vaccines were developed so fast," *Nebraska Medicine*, August 19, 2021, https://www.nebraskamed.com/COVID/were-the-covid-19-vaccines-rushed.

5 "COVID-19 Vaccine Development: Behind the Scenes" NIH: National Institute of Allergy and Infectious Diseases, accessed October 16, 2022, https://covid19.nih.gov/news-and-stories/vaccine-development.

6 NIH, "COVID-19 Vaccine Development."

7   Samuel Chamberlain, "Fauci admits 'modest' NIH funding of Wuhan lab but denies 'gain of function'," *New York Post*, May 25, 2021, https://nypost.com/2021/05/25/fauci-admits-nih-funding-of-wuhan-lab-denies-gain-of-function/.

8   Samuel Chamberlain, "Pentagon gave millions to EcoHealth Alliance for weapons of research program," *New York Post*, July 1, 2021, https://nypost.com/2021/07/01/pentagon-gave-millions-to-ecohealth-alliance-for-wuhan-lab/.

9   Edward C. Homes, et al., "The origins of SARS-CoV-2: A critical review," Cell 184, (September 16, 2021): 4853.

10 Holmes, et al., "The origins of SARS-CoV-2," 4853.

11 Holmes, et al., "The origins of SARS-CoV-2," 4853.

12 Rose Williams, "Wuhan whistleblower: Not 'one bit of evidence' that COVID-19 'naturally emerged," *Alpha News*, October 14, 2022, https://alphanews.org/wuhan-whistleblower-not-one-bit-of-evidence-that-covid-19-naturally-emerged/.

13 U.S. Department of Health and Human Services, National Institute of Health, Letter to Congressman Comer, Lawrence A. Tabak, (Bethesda, Maryland, October 20, 2021), https://int.nyt.com/data/documenttools/nih-eco-health-alliance-letter/512f5ee70ce9c67c/full.pdf, (letter to Congressman Comer).

14 Ronn Blitzer, "NIH gives new grant to EcoHealth Alliance to research bat coronaviruses, despite concerns over Wuhan lab link," Fox News, October 3, 2022, https://www.foxnews.com/politics/nih-gives-new-grant-ecohealth-alliance-research-bat-coronaviruses-despite-concerns-wuhan-lab-link.

15 Jesse O'Neill, "Critics rip Dr. Anthony Fauci after latest Wuhan lab leak findings," *New York Post*, February 27, 2023, https://nypost.com/2023/02/27/anthony-faucis-early-refutal-of-wuhan-lab-leak-under-renewed-criticism/.

16 "Vaccine Development—101," Food and Drug Administration, accessed 17 Oct 2022, https://www.fda.gov/vaccines-blood-biologics/development-approval-process-cber/vaccine-development-101.

17 "Understanding How COVID-19 Vaccines Work," Centers for Disease Control and Prevention, September 16, 2022, https://www.cdc.gov/coronavirus/2019-ncov/vaccines/distributing/steps-ensure-safety.html.

18 "ClinicalTrials.gov Background," ClinicalTrials.gov, accessed 17 Oct 2022, https://clinicaltrials.gov/ct2/about-site/background.

19 "Study to Describe the Safety, Tolerability, Immunogenicity, and Efficacy of RNA vaccine candidates against COVID-19 in Healthy Individuals," ClinicalTrials.Gov, accessed October 17, 2022, https://clinicaltrials.gov/ct2/show/NCT04368728.

20 "COVID Data Tracker," Centers for Disease Control and Prevention, October 17, 2022, https://covid.cdc.gov/covid-data-tracker/#vaccinations_vacc-people-additional-dose-totalpop.

21 Sameer S. Chopra, "Industry Funding of Clinical Trials: Benefit or Bias?" *JAMA* 290, no. 1, (July 2003): 113. https://jamanetwork.com/journals/jama/fullarticle/196846.

22 Ben Goldacre, "Trial sans Error: How Pharma-Funded Research Cherry-Picks Positive Results [Excerpt]," *Scientific American,* February 13, 2013, https://www.scientificamerican.com/article/trial-sans-error-how-pharma-funded-research-cherry-picks-positive-results/.

23 Jackson v. Ventavia Research Group, Relator Brook Jackson's Amended Complaint, Case No. 1:12-cv-00008-MJT, (E.D. Texas. Feb 22, 2022).

24 Robert G. Evans, "Tough on Crime? Pfizer and the CIHR," *Healthcare Policy* 5, no.4 (May 2010): 16, https://www.ncbi.nlm.nih.gov/pmc/articles/PMC2875889/.

25 Evans, "Tough on Crime," 21.

26 Olivier J. Wouters, "Lobbying Expenditures and Campaign Contributions by Pharmaceutical and Health Product Industry in the United States, 1999-2018," *JAMA Intern. Med* 180, no.5 (May 2020): 688, https://jamanetwork.com/journals /jamainternalmedicine/fullarticle/2762509.

Chapter 5

1 "From Thomas Jefferson to Henry Tazewell, 13 September 1795," Founders Online, National Archives, https://founders.archives.gov/documents/Jefferson/01-28-02-0368. [Original source: The Papers of Thomas Jefferson, vol. 28, 1 January 1794–29 February 1796, ed. John Catanzariti (Princeton: Princeton University Press, 2000), p. 466.

2 U.S. Department of Defense, Fact Sheet:2019 Novel Coronavirus (2019-nCOV), https://media.defense.gov/2020/Jan/31/2002242466/-1/-1/1/CORONAVIRUS -FACT-SHEET.PDF.

3 U.S. Department of Defense, "Military Personnel Guidance for Department of Defense Components in Responding to Coronavirus Disease 2019," Virginia S. Pendrod (Washington, D.C. March 23, 2020). https://media.defense.gov/2020/Mar/26 /2002270634/-1/-1/1/Military-Personnel-Guidance-for-Department-of-Defense -Components-in-Responding-to-Coronavirus-Disease-2019.pdf.

4 U.S. Department of Defense, "Guidance for Commanders on Risk-Based Changing of Health Protection Condition Levels During the Coronavirus Disease 2019 Pandemic," Mark Esper (Washington, D.C. May 19, 2020). https://media.defense.gov/2020 /May/20/2002303429/-1/-1/1/GUIDANCE-FOR-COMMANDERS-ON-RISK -BASED-CHANGING-OF-HPCON-DURING-COVID-19.PDF.

5 U.S. Department of Defense, Message to the Force—COVID-19 Response, Mark Esper (Washington, D.C. March 27, 2020). https://www.whs.mil/Portals/75 /Coronavirus/MESSAGE%20TO%20THE%20FORCE%20-%20 COVID-19%20RESPONSE%20OSD003399-20%20FOD%20Final .pdf?ver=2020-03-30-114139-247.

6 U.S. Department of Defense, Message to the Force—COVID-19 Response, Mark Esper.

7 The White House, "Remarks by President Trump on Vaccine Development," (Washington D.C., May 15, 2020), https://trumpwhitehouse.archives.gov/briefings -statements/remarks-president-trump-vaccine-development/.

8 Mark T. Esper, *A Sacred Oath: Memoirs of a Secretary of Defense During Extraordinary Times* (New York: HarperCollins, 2022), 288.

9 Esper, *A Sacred Oath*, 288.

10 Esper, *A Sacred Oath*, 276.

11 Charlie Spiering, "Donald Trump Asked About Coronavirus Questions Surrounding Wuhan Virology Lab," Breitbart, April 15, 2020, https://www.breitbart.com /politics/2020/04/15/donald-trump-asked-about-coronavirus-questions-surrounding -wuhan-virology-lab/.

12   The White House, "Remarks by President Trump on Vaccine Development," (Washington D.C., May 15, 2020), https://trumpwhitehouse.archives.gov/briefings -statements/remarks-president-trump-vaccine-development/.

13   U.S. Department of Defense, Commander U.S. Fleet Forces Command, NAVNORTH FRAGORD 20-024.013 In Response to Coronavirus, (Norfolk, Virginia, 2020), https://nps.edu/documents/106660594/120747831/NAVNORTH-FRAGORD-20 -024.013-IN-RESPONSE-TO-CORONAVIRUS.pdf/3497610b-3dd0-d7cf-eabf -f5980506555b?t=1593044733149.

14   U.S. Department of Defense, Defense Media Activity, "DoD Announces COVID-19 Vaccine Distribution Plan" (Washington, D.C. December 9, 2020), https://www.defense .gov/News/Releases/Release/Article/2440556/dod-announces-covid-19-vaccine -distribution-plan/.

15   Bruce Gillingham, Surgeon General Email message to subordinates, "Subject: We Are At War—Are You Protecting Your Community? (UNCLASSIFIED)," February 26, 2021. https://www.docdroid.com/Tltisw1/20210226redactednavy-surgeon-general-email -pdf

16   U.S. Department of Defense, Message to the Force, Lloyd Austin, (Washington, D.C. Mar 4, 2021). https://media.defense.gov/2021/Mar/04/2002593656/-1/-1/0 /SECRETARY-LLOYD-J-AUSTIN-III-MESSAGE-TO-THE-FORCE.PDF.

17   U.S. Department of Defense, Commander U.S. Fleet Forces Command, NAVNORTH FRAGORD 20-024.013 In Response to Coronavirus, (Norfolk, Virginia, 2020), https://nps.edu/documents/106660594/120747831/NAVNORTH-FRAGORD-20 -024.013-IN-RESPONSE-TO-CORONAVIRUS.pdf/3497610b-3dd0-d7cf-eabf -f5980506555b?t=1593044733149.

18   The White House, "Fact Sheet: President Biden to Announce New Actions to Get More Americans Vaccinated and Slow the Spread of the Delta Variant," (Washington D.C., July 29, 2021), https://www.whitehouse.gov/briefing-room/statements-releases/2021 /07/29/fact-sheet-president-biden-to-announce-new-actions-to-get-more-americans -vaccinated-and-slow-the-spread-of-the-delta-variant/.

19   U.S. Department of Defense, Message to the Force, Lloyd Austin, (Washington, D.C. August 9, 2021). https://media.defense.gov/2021/Aug/09/2002826254/-1/-1/0 /MESSAGE-TO-THE-FORCE-MEMO-VACCINE.PDF.

20   U.S. Department of Defense, Message to the Force, Lloyd Austin.

21   "Revolver Exclusive: Navy Commander Warns of National Security Threat from Mandatory Vaccination," Revolver News, August 15, 2021, https://www.revolver.news /2021/08/navy-commander-warns-national-security-threat-from-mandatory -vaccination/.

22   "Revolver Exclusive," Revolver News, August 15, 2022.

23   "Revolver Exclusive," Revolver News, August 15, 2022.

24   U.S. Department of Defense, Mandatory Coronavirus Disease 2019 Vaccination of Department of Defense Service Members, Lloyd Austin, (Washington, D.C. August 24, 2021). https://media.defense.gov/2021/Aug/25/2002838826/-1/-1/0 /MEMORANDUM-FOR-MANDATORY-CORONAVIRUS-DISEASE-2019 -VACCINATION-OF-DEPARTMENT-OF-DEFENSE-SERVICE-MEMBERS.PDF . emphasis added

25   "COVID-19 Vaccination Administrative Counseling/Warning," September 3, 2021, https://www.docdroid.com/VjnJUCf/cdr-robert-a-green-jr-order-to-be-vaccinated-pdf.

Chapter 6

1   "Amendments to the Constitution, [8 June] 1789," Founders Online, National Archives, https://founders.archives.gov/documents/Madison/01-12-02-0126. [Original source: *The Papers of James Madison,* vol. 12, 2 March 1789–20 January 1790 and supplement 24 October 1775–24 January 1789, ed. Charles F. Hobson and Robert A. Rutland. Charlottesville: University Press of Virginia, 1979, pp. 196–210.]

2   Joseph Loconte, "James Madison and Religious Liberty," The Heritage Foundation Executive Memorandum, March 16, 2001, https://www.heritage.org/political-process /report/james-madison-and-religious-liberty.

3   Madeline Osburn, "3 Sickening Truths About Aborted Fetus Trafficking We Learned From The Daleiden Hearings," *The Federalist,* September, 25, 2019, https://thefederalist. com/2019/09/25/3-sickening-truths-about-aborted-fetus-trafficking-we-learned-from -the-daleiden-hearings/.

4   Negar Motayagheni, "Modified Langendorff technique for mouse heart cannulation: Improved heart quality and decreased risk of ischemia," *MethodsX* 4, (2017): 510, https://doi.org/10.1016/j.mex.2017.11.004.

5   Feng Lan, et al., "Safe Genetic Modification of Cardiac Stem Cells Using a Site-Specific Integration Technique," *Circulation* 126, no.11 (Sept 2012): 20–28 https://www .ahajournals.org/doi/10.1161/CIRCULATIONAHA.111.084913.

6   Chung-E Tseng, et al.," mRNA and Protein Expression of SSA/Ro and SSB/La in Human Fetal Cardiac Myocytes Cultured Using a Novel Application of the Langendorff Procedure," *Pediatric Research* 45, (1999): 260–269 https://www.nature.com/articles /pr199910z.

7   Tseng, "mRNA and Protein Expression," *Pediatric Research.*

8   Nuremberg Code, United States Holocaust Memorial Museum, accessed November 1, 2022, https://www.ushmm.org/information/exhibitions/online-exhibitions/special -focus/doctors-trial/nuremberg-code,

9   Richard Sisk, "Navy Drops Ban on Attending Indoor Religious Services Off-Base," Military.com, July 10, 2020, https://www.military.com/daily-news/2020/07/10/navy -drops-ban-attending-indoor-religious-services-off-base.html.

10 U.S. Department of Defense, "Clarification of Guidance Related to Attendance at Religious Services," Gregory J Slavonic, accessed November 1, 2022, https://www .lifesitenews.com/wp-content/uploads/2021/03/SECDEF_MEMO.pdf.

Chapter 7

1   "From Benjamin Franklin to Jane Mecom, 12 February 1756," Founders Online, National Archives, https://founders.archives.gov/documents/Franklin/01-06-02-0167. [Original source: *The Papers of Benjamin Franklin,* vol. 6, April 1, 1755, through September 30, 1756, ed. Leonard W. Labaree (New Haven and London: Yale University Press, 1963), pp. 400–401.

2   Joy Neal Kidney, "Tap Code by Col. Carlyle "Smitty" Harris," Book review on JoyNealKidney.com, January 22, 2020, https://joynealkidney.com/2020/01/22/tap-code -by-col-carlyle-smitty-harris/.

3   Patrick Byrne, "Affidavit of LTC. Theresa Long M.D. in Support of a Motion for a Preliminary Injunction," *Guardians of Medical Choice,* accessed November 3, 2022,

https://guardiansofmedicalchoice.com/wp-content/uploads/2021/09/Affidavit
-Theresa-Long.pdf.

4   Jason Morgan, "I am 100% Convinced That It's Biological Warfare": Lieutenant Colonel
Pete Chamber' Special Interview with The Remnant," *The Remnant Newspaper,* February
16, 2023, https://remnantnewspaper.com/web/index.php/articles/item/6394-i-am-100
-convinced-that-it-s-biological-warfare-lieutenant-colonel-ret-pete-chambers-special
-interview-with-the-remnant.

5   Lieutenant Colonel Pete C. Chambers, "SARS-CoV-2 Investigational Vaccine "Critical
Thinking"/Informed Consent," Brief to Task Force Salerno, accessed February 17,
2023, https://img1.wsimg.com/blobby/go/058ad340-73c5-4f3d-af4f-8df4795d5196
/COVID%20BRIEF%202021_12.pdf.

6   Jason Morgan, "I am 100% Convinced That It's Biological Warfare": Lieutenant Colonel
Pete Chamber' Special Interview with The Remnant," *The Remnant Newspaper,* February
16, 2023, https://remnantnewspaper.com/web/index.php/articles/item/6394-i-am-100
-convinced-that-it-s-biological-warfare-lieutenant-colonel-ret-pete-chambers-special
-interview-with-the-remnant.

7   Morgan, "I am 100% Convinced That It's Biological Warfare."

8   U.S. Department of Defense, Chief of Naval Operations, COVID-19 Consolidated
Disposition Authority (CCDA), William Lescher and John B. Nowell, (Washington
D.C., October 13, 2021), https://www.mynavyhr.navy.mil/Portals/55/Messages
/NAVADMIN/NAV2021/NAV21225.txt?ver=EfkG2psijI2X0IEKSId_5w%3d%3d.

9   U.S. Department of Defense, Secretary of the Navy, 2021-2022 Department of Navy
Mandatory COVID-19 Vaccination Policy, Carlos Del Toro, (Washington D.C., August
30, 2021), https://www.mynavyhr.navy.mil/Portals/55/Messages/ALNAV/ALN2021
/ALN21062.txt?ver=Vbl_3soAE1K4DhYwqjSGLw%3d%3d.

Chapter 8

1   "John Adams to Abigail Adams, 27 April 1777," Founders Online, National Archives,
https://founders.archives.gov/documents/Adams/04-02-02-0170. [Original source:
The Adams Papers, Adams Family Correspondence, vol. 2, June 1776–March 1778,
ed. L. H. Butterfield (Cambridge, MA: Harvard University Press, 1963), pp. 224–226.

2   U.S. Department of Defense, Secretary of the Navy, 2021-2022 Department of Navy
Mandatory COVID-19 Vaccination Policy, Carlos Del Toro.

3   U.S. Department of Defense, Chief of Naval Operations, 2021-2022 Navy Mandatory
COVID-19 Vaccination and Reporting Policy, W. R. Merz, (Washington D.C.,
August 31, 2021), https://www.mynavyhr.navy.mil/Portals/55/Messages/NAVADMIN
/NAV2021/NAV21190.txt?ver=mG6Zday9ICjIOVsV4HyZEw%3d%3d.

4   U.S. Congress, House, Committee on Government Reform, The Department of
Defense Anthrax Vaccine Immunization Program: Unproven Force Protection, 106th
Cong., 2d Sess., 2000. H. Rep. 106–556, 2 https://www.govinfo.gov/content/pkg
/CRPT-106hrpt556/html/CRPT-106hrpt556.htm.

5   "Personal Protective Equipment EUAs," U.S. Food and Drug Administration, accessed
November 10, 2022, https://www.fda.gov/medical-devices/coronavirus-disease-2019
-covid-19-emergency-use-authorizations-medical-devices/personal-protective
-equipment-euas.

6   In Vitro Diagnostics EUAs – Antigen Diagnostic Tests for SARS-CoV-2," U.S. Food
    and Drug Administration, accessed November 10, 2022, https://www.fda.gov/medical
    -devices/coronavirus-disease-2019-covid-19-emergency-use-authorizations-medical
    -devices/in-vitro-diagnostics-euas-antigen-diagnostic-tests-sars-cov-2.

7   "NEWS: DailyMed Announcements," National Institute of Health: National Library
    of Medicine, accessed November 10, 2022. https://dailymed.nlm.nih.gov/dailymed
    /dailymed-announcements-details.cfm?date=2021-09-13.

8   Lee Vliet, "Dept. of Defense Violations of Law," Truth for Health Foundation, accessed
    November 10, 2022, Encl. 8, https://www.truthforhealth.org/2022/09/dept-of-defense
    -violations-of-law/.

9   Lee Vliet, "Dept. of Defense Violations of Law," Truth for Health Foundation, accessed
    November 10, 2022, Encl. 9, https://www.truthforhealth.org/2022/09/dept-of-defense
    -violations-of-law/.

10 I Lee Vliet, "Dept. of Defense Violations of Law," Truth for Health Foundation, accessed
    November 10, 2022, Encl. 10, https://www.truthforhealth.org/2022/09/dept-of
    -defense-violations-of-law/.

11 "Q&A for Comirnaty (COVID-19 Vaccine mRNA)," U.S. Food and Drug
    Administration, accessed November 10, 2022, https://www.fda.gov/vaccines-blood
    -biologics/qa-comirnaty-covid-19-vaccine-mrna.

12 "Purple Book Database of Licensed Biological Products," U.S. Food and Drug
    Administration, accessed November 10, 2022, https://purplebooksearch.fda.gov
    /results?query=COVID-19%20Vaccine,%20mRNA&title=Comirnaty. (Simple Search
    Results for: Comirnaty).

13 U.S. Department of Health and Human Services, Food and Drug Administration,
    Emergency Use Authorization Reissuance Letter, Peter Marks, (Maryland, December 8,
    2022), 20, https://www.fda.gov/media/150386/download.

14 Tolkien, J. R. R., The Silmarillion, ed. Christopher Tolkien, (New York: Houghton
    Mifflin Company, 2004).

Chapter 9

1   "From George Washington to Major General Nathanael Greene, 5 May 1778,"
    Founders Online, National Archives, https://founders.archives.gov/documents
    /Washington/03-15-02-0040. (Draft written in Alexander Hamilton's handwriting).
    [Original source: The Papers of George Washington, Revolutionary War Series, vol. 15,
    May–June 1778, ed. Edward G. Lengel (Charlottesville: University of Virginia Press,
    2006), pp. 41–42.

Chapter 10

1   "To George Washington from Major General Benedict Arnold, 6 August 1780,"
    Founders Online, National Archives, https://founders.archives.gov/documents
    /Washington/03-27-02-0401. [Original source: The Papers of George Washington,
    Revolutionary War Series, vol. 27, 5 July–27 August 1780, ed. Benjamin L. Huggins
    (Charlottesville: University of Virginia Press, 2019), p. 444.

2   David Barno and Nora Bensahel, "The Increasingly Dangerous Politicization of the U.S. Military," War on the Rocks, June 18, 2019, https://warontherocks.com/2019/06/the -increasingly-dangerous-politicization-of-the-u-s-military/.

3   Vice Admiral William "Dean" Lee, "USCG (Ret) Vice Admiral Lee Calls out Coast Guard Abuses of Expelling Cadets," Interview on The Whistleblower Report, America Out Loud, October 6, 2022, 35:54, https://www.americaoutloud.com/uscg-ret-vice -admiral-lee-calls-out-coast-guard-abuses-of-expelling-cadets/.

4   Grant Atkinson, "General Mark Milley Vowed to 'Fight' Trump 'From the Inside' in 2020: Report," The Western Journal, August 9, 2022, https://www.westernjournal.com /general-mark-milley-vowed-fight-trump-inside-2020-report/.

5   Atkinson, "General Mark Milley Vowed to 'Fight' Trump."

6   Bob Woodward and Robert Costa, Peril (New York: Simon & Schuster, 2021), 129.

7   Woodward and Costa, Peril, 129.

8   Zachary Cohen, Oren Liebermann, and Ellie Kaufman, "Milley Defends Trump-era calls to Chinese counterpart in Congressional Afghanistan hearing," CNN, September 28, 2021, https://www.cnn.com/2021/09/28/politics/afghanistan-hearing-milley-austin -mckenzie/index.html.

9   Cohen, Liebermann, and Kaufman, "Milly Defends Trump-era calls."

10 Nathaniel Philbrick, "Why Benedict Arnold Turned Traitor Against the American Revolution: The story behind the most famous betrayal in U.S. history shows the complicated politics of the nation's earliest days," Smithsonian Magazine, May 2016, https://www.smithsonianmag.com/history/benedict-arnold-turned-traitor-american -revolution-180958786/.

11 "From George Washington to Major General Benedict Arnold, 3 August 1780," Founders Online, National Archives, https://founders.archives.gov/documents /Washington/03-27-02-0361. [Original source: The Papers of George Washington, Revolutionary War Series, vol. 27, 5 July–27 August 1780, ed. Benjamin L. Huggins.

12 "To George Washington from Major General Benedict Arnold, 6 August 1780," Founders Online, National Archives, https://founders.archives.gov/documents /Washington/03-27-02-0401. [Original source: The Papers of George Washington, Revolutionary War Series, vol. 27, 5 July–27 August 1780, ed. Benjamin L. Huggins (Charlottesville: University of Virginia Press, 2019), p. 444.

Chapter 11

1   "Abigail Adams to John Quincy Adams, 21 July 1786," Founders Online, National Archives, https://founders.archives.gov/documents/Adams/04-07-02-0102. [Original source: The Adams Papers, Adams Family Correspondence, vol. 7, January 1786–February 1787, ed. C. James Taylor, Margaret A. Hogan, Celeste Walker, Anne Decker Cecere, Gregg L. Lint, Hobson Woodward, and Mary T. Claffey (Cambridge, MA: Harvard University Press, 2005), pp. 274–277.

2   U.S. Navy SEALs 1-26 v. Biden, Case 4:21-cv-01236-O, Doc 1, Court Listener, Free Law Project, (U.S. District Court N.D. Texas, November 9, 2021) at 1, https://www .courtlistener.com/docket/60824061/1/us-navy-seals-1-26-v-biden/.

3   U.S. Navy SEALs 1-26 v. Biden, Case 4:21-cv-01236-O, Doc 1, Court Listener, Free Law Project, (U.S. District Court N.D. Texas, November 9, 2021) at 16, https://www .courtlistener.com/docket/60824061/1/us-navy-seals-1-26-v-biden/.

4   Kelly Laco, "Coast Guard used 'digital tool' to more efficiently mass deny religious vax exemptions, Republicans allege," Fox News, October 18, 2022, https://www.foxnews .com/politics/coast-guard-digital-tool-efficiently-mass-deny-religious-vax-exemptions -republicans-allege.

5   Doster v. Kendall, Case 1:22-cv-00084, Doc 1, (U.S. District Court S.D. Ohio, February 16, 2022) at 13, https://www.sirillp.com/wp-content/uploads/2022/07/001 -VERIFIED-CLASS-ACTION-COMPLAINT-FOR-DECLARATORY -JUDGMENT-AND-INJUNCTIVE-RELIEF.pdf.

6   Doster v. Kendall, Case 1:22-cv-00084, Doc 13, (U.S. District Court S.D. Ohio, February 22, 2022) at 5, https://www.sirillp.com/wp-content/uploads/2022/07/013 -EMERGENCY-MOTION-FOR-TRO-MOTION-FOR-PRELIMINARY -INJUNCTION-FILED-BY-PLAINTIFFS.pdf.

7   U.S. Navy SEALs 1-26 v. Biden, Case 4:21-cv-01236-O, Doc 66, Court Listener, Free Law Project, (U.S. District Court N.D. Texas, January 3, 2022), https://www.courtlistener .com/docket/60824061/66/us-navy-seals-1-26-v-biden/, accessed 15 Nov 22

8   U.S. Navy SEALs 1-26 v. Biden, Case 4:21-cv-01236-O, Doc 66, Court Listener, Free Law Project, (U.S. District Court N.D. Texas, January 3, 2022) at 1, https://www .courtlistener.com/docket/60824061/66/us-navy-seals-1-26-v-biden/.

9   U.S. Navy SEALs 1-26 v. Biden, Case 4:21-cv-01236-O, Doc 66, Court Listener, Free Law Project, (U.S. District Court N.D. Texas, January 3, 2022) at 24, https://www .courtlistener.com/docket/60824061/66/us-navy-seals-1-26-v-biden/.

10 U.S. Navy SEALs 1-26 v. Biden, No.22-10077, Document 00516435036, (U.S. Court of Appeals, Fifth Cir., August 16, 2022) at 35.

Chapter 12

1   "Abigail Adams to John Adams, 7 September 1776," Founders Online, National Archives, https://founders.archives.gov/documents/Adams/04-02-02-0079. [Original source: *The Adams Papers, Adams Family Correspondence*, vol. 2, June 1776–March 1778, ed. L. H. Butterfield (Cambridge, MA: Harvard University Press, 1963), pp. 121–124.

2   U.S. Department of Defense, Under Secretary of Defense, Force Health Protection Guidance (Supplement 23) – "Department of Defense Guidance for Coronavirus Disease 2019 Vaccination Attestation and Screening Testing for Unvaccinated Personnel," Gilbert R. Cisneros Jr. (Washington, D.C. September 7, 2021). https://media .defense.gov/2021/Sep/08/2002849177/-1/-1/0/FORCE-HEALTH-PROTECTION -GUIDANCE-SUPPLEMENT%2023-DEPARTMENT-OF-DEFENSE -GUIDANCE-FOR-CORONAVIRUS-DISEASE-2019-VACCINATION -ATTESTATION-AND-SCREENING-TESTING-FOR-UNVACCINATED -PERSONNEL.PDF.

3   U.S. Department of Defense, Under Secretary of Defense, Force Health Protection Guidance (Supplement 23) "Department of Defense Guidance for Coronavirus Disease 2019 Vaccination Attestation, Screening Testing, and Vaccination Verification," Gilbert R. Cisneros Jr., (Washington, D.C. October 18, 2021). https://media.defense.gov/2021 /Oct/18/2002875550/-1/-1/1/FORCE-HEALTH-PROTECTION-GUIDANCE -SUPPLEMENT%2023-REVISION-1-DEPARTMENT-OF-DEFENSE -GUIDANCE-FOR-CORONAVIRUS-DISEASE-2019-VACCINATION -ATTESTATION-SCREENING-TESTING-AND-VACCINATION -VERIFICATION.PDF, accessed 16 Nov 2022

4   U.S. Department of Defense, Chief of Naval Operations, "Required COVID-19 Testing for Unvaccinated Service Members," W. R. Merz (Washington D.C., November 24, 2021), https://www.mynavyhr.navy.mil/Portals/55/Messages/NAVADMIN/NAV2021 /NAV21268.txt?ver=LnSPO7lvOEoXXNd0_mCBRg%3d%3d.

5   Leah Anaya, "Exposing Military Corruption: Army Captain Put Through Mental Health Evaluation for Filing Complaint Against General [VIDEO]," Red Voice Media, April 5, 2022, https://www.redvoicemedia.com/2022/04/exposing-military-corruption-army -captain-put-through-mental-health-evaluation-for-filing-complaint-against-general -video/.

6   U.S. Department of Defense, Department of the Army, Memorandum for the Record, CPT Ritter HIPAA Complaint Additional Information, Seth Ritter, August 6, 2022, pg 2.

7   "Army Whistleblower Speaks Out – Retaliatory Weaponization of Psychiatry," Interview with Captain Seth Ritter on Truth for Health . . . The Rest of the Story, America Out Loud, April 17, 2022, 30:25, https://www.americaoutloud.com/army-whistleblower -speaks-out-retaliatory-weaponization-of-psychiatry/.

8   David Willson, Attorney at Law, Disabled Rights Advocates, "Response to Referral of Report of Investigation ICO CPT Seth Ritter, USA," Memorandum for Commander, MCoE, Fort Benning, Georgia, 22 Aug 2022, at 10.

9   David Willson, Attorney at Law, Disabled Rights Advocates, "Response to Referral of Report of Investigation ICO CPT Seth Ritter, USA," Memorandum for Commander, MCoE, Fort Benning, Georgia, 22 Aug 2022, 10–11.

Chapter 13

1   "Adams' "Abstract of the Argument": Ca. April 1761," Founders Online, National Archives, https://founders.archives.gov/documents/Adams/05-02-02-0006-0002-0003. [Original source: *The Adams Papers, Legal Papers of John Adams,* vol. 2, Cases 31–62, ed. L. Kinvin Wroth and Hiller B. Zobel (Cambridge, MA: Harvard University Press, 1965), pp. 134–144.

2   United States v. Clay, Caselaw Access Project, Harvard Law School, 1 C.M.A. 74, 1 C.M.R. 74, (1 U.S. Court of Military Appeals, November 27, 1951) at 77–78, https:// cite.case.law/cma/1/74/.

3   United States v Booker, Caselaw Access Project, Harvard Law School, 5 M.J. 238, (U.S. Court of Military Appeals, Oct 11, 1977) at 244, https://cite.case.law/mj/5/238/.

4   Lolita C. Baldor, "Navy Commander Fired After Refusing to Get COVID Vaccine," *U.S. News & World Report,* December 10, 2021, https://www.usnews.com/news/politics /articles/2021-12-10/navy-commander-fired-over-vaccine-refusal.

5   Lucian Kins, "CDR Kins Appeal of Non-Judicial Punishment," Appeal Submitted to Commander Naval Surface Forces Atlantic, January 4, 2022, 3.

6   Kins, "CDR Kins Appeal," 3.

7   Kins, "CDR Kins Appeal," 2.

8   Kins, "CDR Kins Appeal," 4.

9   United States v. Valead, Caselaw Access Project, Harvard Law School, 32 M.J. 122, (U.S. Court of Military Appeals, Mar 19, 1991) at 127-128, https://cite.case.law/mj/32 /122/. (Everett, S.J. concurring).

[10] United States v. Edwards, Caselaw Access Project, Harvard Law School, 46 M.J. 41, (U.S. Court of Appeals for the Armed Forces, Feb 28, 1997) at 45, https://cite.case.law /mj/46/41/.

[11] U.S. Department of Defense, Department of the Army, Army Public Health Center, DD FORM 458 Charge Sheet, Special Court-Martial Convened by order of Major General Robert L. Edmonson II, Signed by Yevgeny S. Vindman, Staff Judge Advocate, January 18, 2022.

[12] Zachary Cohen, Jake Tapper, and Paul LeBlanc, "White House Ukraine Expert to testify he reported concerns about Trump-Zelensky call," CNN, October 29, 2019, https:// www.cnn.com/2019/10/28/politics/alexander-vindman-nsc-impeachment-testimony /index.html.

[13] Alexander Vindman, *Here, Right Matters: An American Story* (New York: HarperCollins, 2021), 196.

[14] U.S. v Bashaw, Defense Motion for Recusal, Motion to Recuse Colonel Yevgeny Vindman for Undue Command Influence, (Army Trial Judiciary, First Judicial Court, May 12, 2022), Appellate Exhibit at 4.

[15] U.S. v Bashaw, Defense Motion for Recusal, Motion to Recuse Colonel Yevgeny Vindman for Undue Command Influence, (Army Trial Judiciary, First Judicial Court, May 12, 2022), Appellate Exhibit at 6.

[16] U.S. v Bashaw, Defense Motion for Recusal, Motion to Recuse Colonel Yevgeny Vindman for Undue Command Influence, (Army Trial Judiciary, First Judicial Court, May 12, 2022), Appellate Exhibit at 5.

[17] U.S. v Bashaw, Defense Motion for Recusal, Motion to Recuse Colonel Yevgeny Vindman for Undue Command Influence, (Army Trial Judiciary, First Judicial Court, May 12, 2022), Appellate Exhibit at 7.

[18] The White House, "Remarks by President Biden on Fighting the COVID-19 Pandemic," (Washington D.C., September 9, 2021), https://www.whitehouse.gov/briefing-room /speeches-remarks/2021/09/09/remarks-by-president-biden-on-fighting-the-covid-19 -pandemic-3/.

[19] U.S. v Bashaw, Certified Record of Trial, Volume II of V, Transcript, (Army Trial Judiciary, First Judicial Court, April 28, 2022) at 100. (emphasis added).

[20] Lt. Col. Yevgeny Vindman and Lt. Col. Daniel Maurer, "What to do About Lt. General (retired) Flynn: Military Justice and Civil-Military Relations Considerations," Just Security, June 11, 2021, https://www.justsecurity.org/76874/what-to-do-about-lt -general-retired-flynn-military-justice-and-civil-military-relations-considerations/.

Chapter 14

[1] "Thomas Jefferson to Elizabeth Trist, 1 June 1815," Founders Online, National Archives, https://founders.archives.gov/documents/Jefferson/03-08-02-0419. [Original source: *The Papers of Thomas Jefferson,* Retirement Series, vol. 8, 1 October 1814 to 31 August 1815, ed. J. Jefferson Looney (Princeton: Princeton University Press, 2011), pp. 515–517.

[2] COVID Data Tracker, "COVID-19 Vaccinations in the United States," accessed January 16, 2023, https://covid.cdc.gov/covid-data-tracker/#vaccinations_vacc-total-admin -rate-total.

3   Lolita C. Baldor, "US Army Misses Recruiting Goal; Other Services Squeak By," October 3, 2022, https://www.military.com/daily-news/2022/10/03/us-army-misses-recruiting -goal-other-services-squeak.html.

4   "Navy Recruiting Command Announces Mission Results for Fiscal Year 2022 and Goals for 2023," U.S. Navy Press Office, October 3, 2022, https://www.navy.mil/Press-Office /News-Stories/Article/3177917/navy-recruiting-command-announces-mission-results -for-fiscal-year-2022-and-goal/.

5   Heather Mongilio, "Navy Predicts Challenging Future Recruiting Environment, On Target to Hit Retention Goals," USNI News, May 16, 2022, https://news.usni .org/2022/05/16/navy-predicts-challenging-future-recruiting-environment-on-target -to-hit-retention-goals.

6   "Waltz Requests Critical Race Theory Materials & Presentations from West Point," Press Release, Congressman Mike Waltz, April 8, 2021, https://waltz.house.gov/news /documentsingle.aspx?DocumentID=486.

7   "Waltz Requests Critical Race Theory Materials & Presentations from West Point," Press Release, Congressman Mike Waltz, April 8, 2021, https://waltz.house.gov/news /documentsingle.aspx?DocumentID=486.

8   U.S. Congress, Senate, Committee on Armed Services, Report 117–130, James M. Inhofe National Defense Authorization Act for Fiscal Year 2023: Report (to Accompany S. 4543), 117th Congress, 2d Session, July 18, 2022, 159–160, https://www.congress .gov/117/crpt/srpt130/CRPT-117srpt130.pdf.

9   The Offices of U.S. Senator Marco Rubio & U.S. Representative Chip Roy, "Woke Warfighters: How Political Ideology is Weakening America's Military," accessed November 27, 2022, https://www.rubio.senate.gov/public/_cache/files/ee1d7a86 -6d0c-4f08-bd15-24e5b28e54b7/3756824FA9C21B819BB97AAB16221530.woke -warfighters-report-3.pdf.

10 Meghann Myers and Leo Shane III, "The vast majority of troops kicked out for COVID vaccine refusal received general discharges," April 27, 2022, https://www.militarytimes .com/news/pentagon-congress/2022/04/27/the-vast-majority-of-troops-kicked-out-for -covid-vaccine-refusal-received-general-discharges/.

11 Andreas Wailzer, "Pentagon drops draconian COVID jab mandate for American soldiers," LifeSite News, January 11, 2023, https://www.lifesitenews.com/news/pentagon-drops -draconian-covid-jab-mandate-for-american-soldiers/.

12 U.S. Navy SEALs 1-26 v. Biden, No.22-10077, Document 00516435036, (U.S. Court of Appeals, Fifth Cir., August 16, 2022) at 63.

13 Hope Hodge Seck, "Active Ships in the US Navy," June 23, 2021, https://www.military .com/navy/us-navy-ships.html.

14 U.S. Navy SEALs 1-26 v. Biden, No.22-10077, Document 00516435036, (U.S. Court of Appeals, Fifth Cir., August 16, 2022) at 68.

15 R. Davis Younts, Press Release, May 21, 2022, https://justthenews.com/sites/default /files/2022-05/Moseley%20Press%20Release.pdf.

16 Younts, Press Release, May 21, 2022.

17 U.S. Navy SEALs 1-26 v. Biden, Case 4:21-cv-01236, Doc 134, Court Listener, Free Law Project, (U.S. District Court N.D. Texas, February 28, 2022), https://www .courtlistener.com/docket/60824061/134/us-navy-seals-1-26-v-biden/.

18 U.S. Navy SEALs 1-26 v. Biden, Case 4:21-cv-01236, Doc 140, Court Listener, Free Law Project, (U.S. District Court N.D. Texas, March 28, 2022) at 24, https://www .courtlistener.com/docket/60824061/140/us-navy-seals-1-26-v-biden/.

19   U.S. Navy SEALs 1-26 v. Biden, Case 4:21-cv-01236, Doc 140, Court Listener, Free Law Project, (U.S. District Court N.D. Texas, March 28, 2022) at 5, https://www .courtlistener.com/docket/60824061/140/us-navy-seals-1-26-v-biden/.

20   U.S. Navy SEALs 1-26 v. Biden, Case 4:21-cv-01236, Doc 140, Court Listener, Free Law Project, (U.S. District Court N.D. Texas, March 28, 2022) at 5, https://www .courtlistener.com/docket/60824061/140/us-navy-seals-1-26-v-biden/.

21   Doster v. Kendall, Case 1:22-cv-00084, Doc 77, (U.S. District Court S.D. Ohio, July 27, 2022) at 2.

22   Colonel Financial Management Officer v. Austin, Case 8:22-cv-01275, Doc 229, (U.S. District Court M.D. Florida, August 18, 2022) at 2.

23   Colonel Financial Management Officer v. Austin, Case 8:22-cv-01275, Doc 229, (U.S. District Court M.D. Florida, August 18, 2022) at 41.

24   Colonel Financial Management Officer v. Austin, Case 8:22-cv-01275, Doc 229, (U.S. District Court M.D. Florida, August 18, 2022) at 45.

Chapter 15

1   "From Benjamin Franklin to Lord Howe, 20 July 1776," Founders Online, National Archives, https://founders.archives.gov/documents/Franklin/01-22-02-0307. [Original source: *The Papers of Benjamin Franklin*, vol. 22, March 23, 1775, through October 27, 1776, ed. William B. Willcox (New Haven and London:: Yale University Press, 1982), pp. 518–521.

2   Elrod v. Burns, 427 U.S. 347 (U.S. Supreme Court, June 25, 1976) at 373.

3   Colonel Financial Management Officer v. Austin, Case 8:22-cv-01275, Doc 229, (U.S. District Court M.D. Florida, August 18, 2022) at 43.

4   Colonel Financial Management Officer v. Austin, Case 8:22-cv-01275, Doc 229, (U.S. District Court M.D. Florida, August 18, 2022) at 42

5   Colonel Financial Management Officer v. Austin, Case 8:22-cv-01275, Doc 229, (U.S. District Court M.D. Florida, August 18, 2022) at 42

6   Department of Defense Pilots Concerned by the COVID-19 Vaccine Mandate, "Written Stories from Injured Service Members: A Message to House and Senate from DoD Pilots," Truth for Health Foundation, accessed 4 December, 2022, https://www .truthforhealth.org/2022/08/written-stories-from-injured-service-members-a-message -to-house-and-senate-from-dod-pilots/.

7   Concerned DoD Pilots, "Written Stories from Injured Service Members," 4.

8   Concerned DoD Pilots, "Written Stories from Injured Service Members," 5.

9   Concerned DoD Pilots, "Written Stories from Injured Service Members," 12.

10   Concerned DoD Pilots, "Written Stories from Injured Service Members," 65.

11   Concerned DoD Pilots, "Written Stories from Injured Service Members," 63.

12   U.S. Department of Defense, Defense Suicide Prevention Office, Department of Defense (DoD) Quarterly Suicide Report (QSR) 3rd Quarter, CY 2022, Liz Clark, (2023), 4.

13   "From Benjamin Franklin to Lord Howe, 20 July 1776," Founders Online, National Archives, https://founders.archives.gov/documents/Franklin/01-22-02-0307. [Original source: *The Papers of Benjamin Franklin*, vol. 22, March 23, 1775, through October 27, 1776, ed. William B. Willcox (New Haven and London:: Yale University Press, 1982), pp. 518–521.

14   "From Benjamin Franklin to Lord Howe, 20 July 1776," Founders Online, pp. 518–521.

Chapter 16

[1]   "From Thomas Jefferson to Albert Gallatin, 24 November 1808," Founders Online, National Archives, https://founders.archives.gov/documents/Jefferson/99-01-02-9148.

[2]   U.S. Department of Defense, Office of the Naval Inspector General Senior Official Investigations Division (IG50), Notification of Case Closure (Case 202106692), August 5, 2022. (This email is enclosure 12 of the Article 1150 Complaint of wrong against Vice Admiral Fuller found at https://www.truthforhealth.org/2022/09/dept-of-defense-violations-of-law/).

[3]   U.S. Department of Defense, Department of the Navy, Assistant Secretary of the Navy, Use of Pfizer-BioNTech Vaccine for Mandatory Vaccination, Robert D. Hogue, September 8, 2021.

[4]   Coker v. Austin, Case 3:21-cv-01211, Doc 31, Court Listener, Free Law Project, (U.S. District Court N.D. Florida, October 21, 2021) at 16, https://www.courtlistener.com/docket/60630202/31/13/coker-v-austin/.

[5]   Coker v. Austin, Case 3:21-cv-01211, Doc 31, Court Listener, Free Law Project, (U.S. District Court N.D. Florida, October 21, 2021) at 16.

[6]   Coker v. Austin, Case 3:21-cv-01211, Doc 31, Court Listener, Free Law Project, (U.S. District Court N.D. Florida, October 21, 2021) at 16.

[7]   Coker v. Austin, Case 3:21-cv-01211, Doc 31, Court Listener, Free Law Project, (U.S. District Court N.D. Florida, October 21, 2021) at 4.

[8]   Coker v. Austin, Case 3:21-cv-01211, Doc 31, Court Listener, Free Law Project, (U.S. District Court N.D. Florida, October 21, 2021) at 4.

[9]   Coker v. Austin, Case 3:21-cv-01211, Doc 31, Court Listener, Free Law Project, (U.S. District Court N.D. Florida, October 21, 2021) at 5.

[10] Coker v. Austin, Case 3:21-cv-01211, Doc 31, Court Listener, Free Law Project, (U.S. District Court N.D. Florida, October 21, 2021) at 5.

[11] Kristina Wong, "Watchdog Warned Lloyd Austin of 'Potential Noncompliance' as DoD Denied Religious Vaccine Exemptions," Breitbart, September 14, 2022, https://www.breitbart.com/politics/2022/09/14/watchdog-warned-lloyd-austin-of-potential-noncompliance-as-dod-denied-religious-vaccine-exemptions/.

[12] Danny, "DODIG Memo to SECDEF Highlights Deliberate Violation of Federal Law within the DOD," *TRMLX*, September 13, 2022, https://trmlx.com/dodig-memo-to-secdef-highlights-deliberate-violation-of-federal-law-within-the-dod/.

Chapter 17

[1]   "Abigail Adams to Mercy Otis Warren, 25 April 1798," Founders Online, National Archives, https://founders.archives.gov/documents/Adams/04-12-02-0272. [Original source: *The Adams Papers, Adams Family Correspondence*, vol. 12, March 1797 – April 1798, ed. Sara Martin, C. James Taylor, Neal E. Millikan, Amanda A. Mathews, Hobson Woodward, Sara B. Sikes, Gregg L. Lint, and Sara Georgini (Cambridge, MA: Harvard University Press, 2015), pp. 526–529.

[2]   Dan Diamond, "Biden's claim that 'pandemic is over' complicates efforts to secure funding," *The Washington Post*, September 19, 2022, https://www.washingtonpost.com/health/2022/09/18/biden-covid-pandemic-over/.

3   U.S. Department of Health & Human Services, Administration for Strategic Preparedness & Response, Renewal of Determination that a Public Health Emergency Exists, Xavier Becerra, January 11, 2023, https://aspr.hhs.gov/legal/PHE/Pages/covid19-11Jan23.aspx.

4   U.S. Department of Health & Human Services, Health Resources & Services Administration, Countermeasure Injury Compensation Program (CICP) Data, data as of December 1, 2022, https://www.hrsa.gov/cicp/cicp-data#table-1.

5   U.S. Department of Health & Human Services, Health Resources & Services Administration, Countermeasure Injury Compensation Program (CICP) Data, data as of December 1, 2022, https://www.hrsa.gov/cicp/cicp-data#table-1.

6   U.S. Department of Defense, Defense Health Agency, DHA Initial Case No: 21-00359 (Other category) Requester's Tracking No 256601, April 20, 2022. (This FOIA request response is enclosure 2 of the August 15, 2022 Whistleblower Report and can be found at https://www.truthforhealth.org/2022/08/whistleblower-report-of-illegal-dod-activity/).

7   Coker v. Austin, Case 3:21-cv-01211, Doc 88, Court Listener, Free Law Project, (U.S. District Court N.D. Florida, May 20, 2022) at 6, https://www.courtlistener.com/docket/60630202/88/1/coker-v-austin/.

8   U.S. Congress, Senate, Senator Ron Johnson, Senate Homeland Security and Governmental Affairs Committee, Declaration of 1LT. Mark C. Bashaw in Support of Senator Ron Johnson Investigation Into the Safety and Efficacy of COVID-19 Vaccines, August 4, 2022, (This declaration is enclosure 9 of the August 15, 2022 Whistleblower Report and can be found at https://www.truthforhealth.org/2022/08/whistleblower-report-of-illegal-dod-activity/).

9   COVID-19 Vaccine Lot. Number and Expiration Date Report, Centers for Disease Control and Prevention, accessed December 17, 2022, https://vaccinecodeset.cdc.gov/LotNumber.

10 U.S. Department of Health & Human Services, Food and Drug Administration, BLA Approval, Mary A Malarkey and Marion F. Gruber, August 23, 2021, https://www.fda.gov/media/151710/download.

11 U.S. Congress, Senate, Senator Ron Johnson, Senate Homeland Security and Governmental Affairs Committee, Declaration of LT Chad Coppin, July 30, 2022, (This declaration is enclosure 11 of the August 15, 2022 Whistleblower Report and can be found athttps://www.truthforhealth.org/2022/08/whistleblower-report-of-illegal-dod-activity/).

12 Megan Brenan, "Americans' Confidence in Major U.S. Institutions Dips," Gallup, July 14, 2021, https://news.gallup.com/poll/352316/americans-confidence-major-institutions-dips.aspx

13 U.S. Congress, Senate, Senator Ron Johnson, Senate Homeland Security and Governmental Affairs Committee, Letter to Secretary Austin, Commissioner Califf, and Director Walensky, August 18, 2022, https://www.ronjohnson.senate.gov/services/files/2266AE5A-027A-42F2-90E1-D4430F32BF37.

14 "The Whistleblower Report," Truth for Health Foundation, accessed December 17, 2022, https://www.truthforhealth.org/blogs-podcast-media-etc/the-whistleblower-report/.

Chapter 18

1   "From John Adams to John Winthrop, 23 June 1776," Founders Online, National Archives, https://founders.archives.gov/documents/Adams/06-04-02-0134. [Original source: *The Adams Papers, Papers of John Adams*, vol. 4, February–August 1776, ed. Robert J. Taylor (Cambridge, MA: Harvard University Press, 1979), pp. 331–333.

2   Ryan Morgan, "Pentagon forcing 'unnecessary' COVID vaccine mandate, says fmr. Marine lawyer who fought anthrax vaccine mandate," *American Military News*, September 3, 2021, https://americanmilitarynews.com/2021/09/pentagon-forcing-unnecessary -covid-vaccine-mandate-says-fmr-marine-lawyer-who-fought-anthrax-vaccine -mandate/.

Chapter 19

1   "General Orders, 4 December 1777," Founders Online, National Archives, https:// founders.archives.gov/documents/Washington/03-12-02-0493. [Original source: The Papers of George Washington, Revolutionary War Series, vol. 12, 26 October 1777–25 December 1777, ed. Frank E. Grizzard, Jr. and David R. Hoth (Charlottesville: University Press of Virginia, 2002), p. 534.

Epilogue

1   "From John Adams to Thomas McKean, 26 November 1815," Founders Online, National Archives, https://founders.archives.gov/documents/Adams/99-02-02-6545.

2   Paul Vicers, "The Air Force's Covid Cure Is Worse Than the Disease," *The American Conservative*, December 16, 2022, https://www.theamericanconservative.com/the-air -force-vaccinated-against-success/.

3   Vicers, "The Air Force's Covid Cure Is Worse Than the Disease."

INDEX

A

abortion, 36, 37
  See Also Langendorff apparatus
Adams, Abigail, 79, 86, 87, 94, 135, 145
Adams, John, 1, 53, 95, 147, 152, 159, 163, 164
Adams, Samuel, 5, 6, 7
Adirim, Terry, 56
American Revolution, 5–7, 76, 164
Aris, Charles, Major General, 48, 90
Arnold, Benedict
  dispatches to Gen Washington before treason, 69, 76
  doubt in America becomes a career opportunism, 76
  membership in the Sons of Liberty, 7
  similarities to our highest ranking leaders, 76
Austin, Lloyd, Secretary of Defense
  discrimination against unvaccinated SVMs, 31
  DoD IG memo admitting RFRA violations, 132
  extremism stand-down order, 2, 106
  mandate of COVID-19 vaccines, 32
  military COVID virus deaths at time of mandate, 123
  prioritizing COVID-19 *above* threat posed by China, 30

B

Bashaw, Mark, Army 1LT, congressional whistleblower
  continued efforts as a whistleblower, 140
  court-martial of, 99–103
Becerra, Xavier, 135
Beckerman, David, Air Force Maj, 48, 142, 144, 150
Berry, Michael, attorney, 80, 83
Bill of Rights, 7, 9, 10, 16, 35
Blackman, Joshua, 12, 13, 15
Bowes, John, Air Force 1LT, 117
Bruns, Thomas, attorney, 111
Buck v. Bell, 12–16, 102
  used by SS Officer as justification for war crimes, 15
  See Also Holmes, Oliver W., Supreme Court Justice

C

Captain Courageous, 64, 68, 122
careerism, 41, 48, 59, 90, 157
  See Also politicization of the military
Carlson, Tucker, 149

CDC, 22, 140, 143

censorship
early pandemic implementation of, 19
ignored by vast majority of media and
journalists, 148
stories of harm hidden by, 115
tech companies enabling domestic
threats through, 147

Chambers, Peter, Dr., Army Lt. Col.
(Ret), 47, 144, 150

Chang, Kristina, Navy LT DDS, 90

Cheek, Jon, Army Lt. Col., 104,
140, 142

China
Austin prioritizing COVID-19 *above*
threat posed by, 30
DoD initial reaction to COVID-19
outbreak in, 27
likely exploitation of pandemic by, 29
US delegation of biosafety at Wuhan
lab to, 21

clinical trials
all government listings of, 22
big pharma funding and
conducting the, 23
CDC spinning claim of no skipped
testing phase, 22
description of each phase of, 22
fraud committed by Pfizer during, 23

Cohen, Robert, Army Judge, 101,
102, 105
See Also politicization of the military

Coker v. Austin, 130, 131, 139

Coker, Benjamin, Navy Chief, 104

*Colonel Financial Management Officer v.
Austin*, 111

Coppin, Chad, USCG LT, 141, 142

Countermeasure Injury Compensation
Program (CICP), 137

court-martial
desperation by leadership to avoid, 51,
96, 99, 109

Fifth and Sixth Amendment rights
afforded by, 96
Gen Milley admitting his own
possibility of, 75
Mark Bashaw trial by, 99–103
Navy's illegal denial of right to
trial by, 97–99, 108
necessity of setting EUA law
precedent at, 96, 99, 109
possibility of jail time for
unvaccinated, 63
right to demand trial by, 95, 97, 162
threatened by leadership, 58, 62,
71, 95
Vindman bragging about the
Bashaw trial by, 103

Crandall, Darse, Vadm, Navy Judge
Advocate General
bragging about role in eroding
readiness, 85
ignored questionable legality in
defense of policy, 66
involvement in unlawful
accommodation denials, 67
obligation to ensure Navy policy was
lawful, 65
politicization of military
enabled by, 85, 96, 109
role in Adm Lescher's court
testimony, 85
See Also politicization of the military

D

Declaration of Independence, 7, 25,
35, 152
Defense Health Agency, 139
Degenkolb, Olivia, Navy
CDR, 49, 90, 104, 142, 144
Del Toro, Carlos, Secretary
of the Navy, 54, 130, 131
Delayed Entry Program, 45, 106

dereliction of duty
  at the highest levels of military
      leadership, 33, 162
  by Adm Crandall and the JAG
      Corps, 67
  committed by Adm Fuller, 127
DODINST 1300.17 (Religious Liberty
      Inst.), 36, 46, 62
domestic threats
  awakening defenders of the
      Constitution, 157
  censorship hiding the
      actions of, 147
  denying medical freedom
      rights, 156
Donahoe, Patrick, Army Major
      General
  allegations of unlawful actions
      committed by, 91
  inappropriate online
      fraternization by, 91
Doster v. Kendall, 111
Dowd, Edward, 150
Duncan, Scott, USMC Lt Col, 63,
      149, 150

E

EcoHealth Alliance, 20, 21
Edmonson, Robert, Army
      Major General, 100
Esper, Mark, 27, 28, 30
  See Also Operation Warp Speed
Extremism Stand-Down, 2, 106
  See Also SASC report on inappropriate
      use of tax funds

F

FDA, 20, 23, 54–57, 130–32,
      138–41, 143
Fifth Amendment, 7, 51, 96,
      108, 109

First Amendment, 39, 69, 81, 116,
      117, 163
First Liberty Institute, 80, 83, 110
Fourteenth Amendment, 8, 37
  See Also abortion
Franklin, Benjamin, 9, 43, 115,
      123, 153
free speech, 36, 69, 144, 149
Fuller, John, Naval Inspector
      General, 126–29, 132–34
Furman, Jay, Navy CDR, 32, 149

G

gain of function research, 20, 21
Gannam, Roger, attorney, 111
Gateway to Freedom Conference, 150
Gilday, Michael, Admiral, 54, 65, 67,
      72, 75, 82, 83
Gillingham, Bruce, Navy Surgeon
      General
  complicit in Adm Grady's apparent
      mutiny, 74
  echoing politicians in proselytizing
      COVID vaccines, 30
  first interchangeability misinformation
      memo, 56, 129
  protected by Adm Fuller in NAV IG
      cover-up, 126
  referenced in Adm Grady's unlawful
      order, 73
Grady, Christopher Admiral
  confirmed by Senate as Vice Chairman
      of JCS, 75, 126
  eagerness to issue vaccine orders, 54
  establishing a virus as a national
      security threat, 30
  first to issue unlawful COVID-19
      vaccine order, 73
  forced by Trump admin to restore
      worship rights, 41
  fulfilling required elements for mutiny
      charge, 74

order to discipline defiers of worship
restrictions, 41
protected by Adm Fuller in NAV IG
cover up, 127
violation of constitutional religious
freedom rights, 40
Graves, Brennan, Air Force
Capt., 161, 162

H

Hacker Stephens, law firm, 83
Hacker, Heather, attorney, 110
Hancock, John, 148
Health Insurance Portability and
Accountability Act
Navy violation of, 98
HHS, 19, 55, 101, 136, 137, 139
Hogue, Robert, 56, 129, 130
Holmes, Oliver W., Supreme Court
Justice
advocated for putting undesirable
infants to death, 13
human rights travesty caused by, 15
intentionally misread *Jacobson*, 13
number of women sterilized by
force because of, 14
quoted by SS officer as defense at
Nuremberg Trials, 15
use of lies in *Buck* to obtain forced
sterilization, 13
Hoppe, Joshua, USMC
Capt., 142, 143
Huff, Andrew Dr., 21

I

informed consent similarity to
Miranda rights, 55, 101
inspector general, 57, 69, 126–30,
132–34, 142
isolation, 44, 45, 47, 51, 121,
142, 161

J

Jackson, Brook, 23
Jacobson v. Massachusetts, 11–12
as precedent and basis for pandemic
jurisprudence, 15
as pretext for compulsory vaccination
principle, 13
falsely recast by Holmes to permit
forced vaccination, 14
JAGs violating ethical obligations, 65,
66, 67, 85, 96, 97
James, Mollie, Dr., 150
Jefferson, Thomas, 27, 33, 105, 114,
125, 134
Johnson, Brandon, attorney, 130
Johnson, Ron, Senator, 140, 143
Josephson, Elizabeth, Navy
CAPT, JAG, 85

K

Karr, Jordan, 151
Kendell, Frank, Secretary of the Air
Force, 82
King, Brandi, Air Force Lt. Col., 144
Kins, Lucian, "Damian,"
Navy CDR, 96, 97, 108
Kupper, Nicholas, Air Force MSgt, 149

L

Laco, Kelly, 149
Langendorff apparatus, 36, 37
See Also abortion
laws
applicability to all Americans, 54,
55, 105
EUA liability protections, 136
EUA product law, 55, 101, 139
interchangeability law, 57, 131
presidential waiver of right to be
informed, 54, 55, 74

proper labeling of FDA-approved
    products, 138, 140
UCMJ Art 15-Vessel
    Exception, 95–98, 109
UCMJ Art 37-Unlawfully influencing
    court action, 103
UCMJ Art 88-Contempt toward
    officials, 71
UCMJ Art 92-Failure to obey
    order, 41, 100, 108
UCMJ Art 94-Mutiny and
    sedition, 73–75
whistleblower protection law, 69
    See Also Religious Freedom
        Restoration Act
Lee, William "Dean", USCG
    Vadm (Ret), 72
Lescher, William, Admiral
    court testimony admitting mandate
        readiness impact, 108
    help from JAG Corps to push political
        agenda, 85
    order to withhold promotion of
        unvaccinated, 49
    willingness to hurt readiness for legal
        win, 84
Liberty Counsel, 111
Loesch, Zachary, USCG Rescue
    Swimmer, 149
Long, Theresa Dr., Army
    Lt. Col., 47, 48

M

MacFarland, Matthew, Federal
    Judge, 111
Macie, Mara, 151
Madison, James, 35
malicious retention
    of Air Force Capt. Brennan
        Graves, 161
    of Air Force Col. JJ McAfee, 142

of Navy Musician First Class Drew
    Stapp, 50, 120
of Navy SEALs leading to a suicide, 121
Malone, Robert, Dr., 150
Marks, Peter, FDA, 130, 131
Martinez, Morgan, Navy ENS, 151
Mast, Richard, attorney, 111
McAfee, John, Air Force Col., 49, 103,
    104, 142
McCullough, Peter, Dr., 150
Meadows, Mark, 76
Meier, John, Navy Radm, 41
Merryday, Steven, Federal Judge, 111,
    113, 116
Miarecki, Sandy, PhD, 150
military medical providers
    abusing service members, 45, 55, 93,
        119, 137, 161
    confirming lack of FDA-approved
        COVID vaccine, 54
    hiding vaccine injuries, 117
    unlawful use of EUA products for
        mandate, 56, 129, 136
Miller, Christopher, 76
Miller, Travis, attorney, 130
Milley, Mark General, Chairman of
    the JCS
    attendance at Op. Warp Speed press
        conference, 28
    informed Chinese he would tip them
        to any attack, 76
    situational similarities to Benedict
        Arnold, 76
    told staff he would fight Trump from
        the inside, 75
moral courage
    flag and general officer lack of, 72
    future of, 80
    requirement to practice the
        virtue of, 156
More, St. Thomas, example of following
    conscience, 63

Morrisette, Michael, Air Force
    SSgt, 120
Moseley, William, Navy LT, 108, 109

N

National Defense Authorization
    Act, 119, 159, 162
National Institute of Health
    (NIH), 19–22, 56
Navy Recruit Training
    Command, 45, 46
Navy Seals 1-26 v. Austin, 86, 107, 110
Navy Special Warfare Command, 80
non-judicial punishment, 95, 97, 99,
    108, 162
Nowell, John Vadm, Chief of Naval
    Personnel
  evidence of breaking the law, 82
  lack of concern for Navy suicides, 122
  order to withhold promotion of
    unvaccinated, 49
  protected by Adm Fuller in NAV IG
    cover up, 128
  protected by O'Donnell in DoD IG
    cover up, 133
  protected from lawsuits by JAG
    Corps, 67
  religious accommodation
    adjudication authority, 39, 64
  religious accommodation disapproval
    SOP, 64–68
  unlawfulness reported to HASC &
    SASC, 84, 94
  unlawfulness resulting in Fed Court
    injunction, 110, 127
Nuremberg Code, 38, 137
Nuremberg Shrug, 59, 126, 141
Nuremberg trials
  head of SS Office quoting Justice
    Holmes as defense, 15
  US officers echoing "just following
    orders" excuse, 59

O

O'Connor, Reed, Federal Judge, 84, 110,
    111, 127, 128
O'Donnell, Sean, Acting DoD IG
  duplicity in admitting then dismissing
    violations, 133
  memo to Austin admitting RFRA
    violations, 132
  similarities to B. Arnold's manufactured
    excuses, 133
  whistleblower exposing apparent fraud
    by, 134
Operation Warp Speed
  altered standards for vaccine
    development, 19
  Esper boasting about, 28
  Esper comparing the Manhattan
    Project to, 29
  First EUA authorization 9 months
    after launch, 20
  launched by the Trump
    administration, 19
  questions about China during press
    conference for, 29
  Rose Garden press conference
    about, 28
  similarities between Manhattan
    Project and, 29
  Trump attempting to justify
    the risk of, 28
  Trump officials unapologetic for
    reckless safety risk, 28

P

Pfizer, 22, 23, 24, 55, 141
  See Also clinical trials
Phelps, J. M., 149
Pike, Robert, Air National Guard Lt.
    Col., 104
Pinnacle, training program for 3-and
    4-star leaders, 72
politicization of the military

Adm Grady prioritizing political
    issues, 54
as demonstrated by Col.
    Vindman, 100–103
careerism deepening the
    crisis of, 56, 58, 113
commanders echoing political talking
    points, 30
enabled by Adm Lescher, 85
enabled by Capt Josephson, 85
enabled by Judge Cohen's falsehood
    at trial, 102
enabled by Vadm Crandall, 66, 85, 96
Gen Aris admits that mandate is about
    politics, 48
harming readiness, 31, 106, 107, 112,
    122, 126, 149
ignoring risk for the sake of political
    agenda, 32
lawfulness of enacting purely political
    policies, 2
political agenda at Recruit Training
    Command, 45
politicization at highest mililtary
    ranks, 72, 90
recent increase in, 71
Vindman calling for a Gen Flynn
    court-martial, 103
Pompeo, Michael, 76
psychological warfare, 119, 122

R

readiness, 45, 62, 84, 106, 107, 108,
    112, 113, 149
recruiting, 32, 45, 105, 110, 112, 125,
    141, 149
Religious Accommodation Request
    Air Force guidance to deny every, 82
    Coast Guard review process, 82
    Navy review process, 66–68, 81–82
religious discrimination, 81, 83,
    107, 113, 116, 141

religious freedom, 12, 15, 35, 40, 63, 81
Religious Freedom Restoration
    Act, 36, 116, 119, 160
Rempfer, Thomas, Air Force
    Col. (Ret), 53, 150
retention   See recruiting
Reyes, Ibrahim, attorney, 130
Ritter, Seth, Army Capt., 92
    unlawful detainment & whistleblower
        reprisal, 91–93
Rocco, Carolyn, Air Force Maj., 104
Roy, Chip, Congressman, 106
Rubio, Marco, Senator, 106
Rush, Benjamin, Dr., 7, 17, 25

S

Saran, Dale, attorney, 53, 150
SASC report on inappropriate use of tax
    funds, 3, 106
Shour, Jonathan, Navy
    Chaplain, 104, 120
Sigoloff, Samuel Dr., Army
    Maj., 47, 150
Siri, Aaron, attorney, 111
Sixth Amendment, 51, 96, 108, 109
Slavonic, Gregory, Acting
    Undersecretary of the Navy, 41
Smith, Grant, Army Capt., 150
Stapp, Andrew, Navy Musician First
    Class, 50, 120
Staver, Mathew, attorney, 111
Stephens, Andrew, attorney, 84
Stewart, Thomas, 150
suicides in the military, 116, 121,
    122, 123
Supreme Court
    Admiral Crandall bragging about
        partial stay by, 85
    Buck ruling undermining
        trust in, 13
    chipping away at individual medical
        freedom rights, 10

denial of First Amendment rights is
   irreparable harm, 116
history of rulings on
   constitutionality, 10–11
Lescher declaration used to partially
   stay injunction, 84
narrow ruling in *Jacobson*, 11–12
responsible for the *Buck* human rights
   travesty, 12–13
wrongfully permitting government
   overreach, 15

T

Tennis, Matthew, Navy LCDR, 39, 64
TerminalCWO, 133
testing for COVID-19
   as a coercive tool, 87, 88
   only EUA testing options available, 88
   punishment for declining, 90, 91, 95,
     97, 100, 108
   right to decline all EUA products, 88
therapeutic proportionality, 38, 89
Third Amendment, 9
Tolkien, John Ronald Rueul, 58, 143
Tomlinson, Carmel, Navy
   CAPT, JAG, 53
Trident Order #12, 81, 84, 107
Truth for Health Foundation, 109,
   117, 143

U

unlawful orders
Alexander Vindman's opinions on, 100
as enabled by Vadm Fuller
   (Nav IG), 132
as issued by Adm Grady, 73
as proven at LT Moseley's separation
   hearing, 109
chain of command requirement to
   screen for, 4
court affirmed obligation to
   resist every, 4

inference of lawfulness, 2, 4, 40
interchangeability falsehood as
   justification for, 57
to murder women and children at My
   Lai, 3

V

Vaccine Adverse Event Reporting Sys.
   (VAERS), 38, 140
Vicars, Paul, Air Force Col., 160
Vindman, Alexander, Army Lt.
   Col., 100
Vindman, Yevgeny, Army Lt.
   Col., JAG, 100, 103
See Also politicization of the military
Vliet, Elizabeth Lee, MD, 144

W

Waltz, Michael, Congressman, 106
Washington, George, 61, 76, 133,
   155, 156
weaponization of psychiatry, 91, 92, 93
Weinstein, Bret, 149
Weist, Christopher, attorney, 111
Westen, John-Henry, 149
White, Candace, Marine Corps Maj.,
   JAG, 53
Wier, Patrick, Navy LCDR,
   JAG, 53, 142
Williams, Darryl, Army Lt. Gen., 106
Wong, Kristina, 149
Wood, L. Todd, 149

Y

Younts, R. Davis, attorney, 109, 149

Z

Zito, Mark, Navy Lieutenant
   Commander, 48, 64